Hard Choices

Climate Change in Canada

Hard Choices: Climate Change in Canada

Edited by Harold Coward and Andrew J. Weaver

Published for The Centre for Studies in Religion and Society
by Wilfrid Laurier University Press

We acknowledge the support of the Canada Council for the Arts for our publishing program. We acknowledge the financial support of the Government of Canada through the Book Publishing Industry Development Program for our publishing activities. We acknowledge the Government of Ontario through the Ontario Media Development Corporation's Ontario Book Initiative.

National Library of Canada Cataloguing in Publication

Hard choices : climate change in Canada / edited by Harold Coward and Andrew J. Weaver.

Includes bibliographical references and index.
ISBN 0-88920-442-X

1. Climatic changes—Canada. I. Coward, Harold G., 1936- II. Weaver, Andrew John, 1961- III. Centre for Studies in Religion and Society.

QC981.8.C5H37 2004 551.6971 C2004-901530-3

Cover design by P.J. Woodland, using a photograph by R.W. Harwood.
Text design by P.J. Woodland.

∞
Printed in Canada

Order from:
Wilfrid Laurier University Press
Waterloo, Ontario, Canada N2L 3C5
www.wlupress.wlu.ca

Contents

Preface

THE AIM OF THIS BOOK is to provide Canadians with an assessment of the implications of climate change for Canada. Leading Canadian scientists, engineers, social scientists, and humanists offer an overview and assessment of climate change and its impacts on Canada from physical, social, technological, economic, political, ethical, and religious perspectives. The large and complex literatures involved from each disciplinary perspective have been summarized and interpreted for the Canadian context, offering a multidisciplinary approach to the challenge of climate change in Canada. Special attention is given to Canada's response to the Kyoto Protocol and to an assessment of the overall adequacy of Kyoto as a response to the global challenge of climate change.

Although humans have always had to cope with the extremes of weather and variations in climate, the advent of modern technology has made the modern challenge of climate change different from the past. Today, our technological and economic strategies give us increased ability to adapt to extreme variations of weather and climate in Canada's various regions, from changes in coastal fisheries to droughts on the prairies and loss of glaciers in the Rocky Mountains: yet the climate change facing us, together with the rapid increase of the world's population and the industrial development of countries like China and India, presents governments and individuals with hard choices now and in the future. Jan Zwicky, a poet and philosopher, introduces us to the issues by telling about her experiences growing up on an Alberta farm and her worries over Canada's changing weather patterns. Following this narrative introduction, Part I provides an overview of the scientific and human challenges of climate change in

Canada—what is happening now and what is going to happen in the future. In the light of this baseline analysis, Part II asks the question, "What can we do?" in response to the climate change challenges. The chapters of Part II focus on terrestrial carbon sinks, technological possibilities, economic responses, regional adaptations, and legal constraints and opportunities. Part III examines the "hard choices" that the challenge of climate change presents to us in terms of Canadian policy developments, developments beyond Kyoto for Canada and the world, and the ethical choices facing us as individuals.

The writing of this book was made possible by funding provided by the Canadian Global Change Program. This book project of the Centre for Studies in Religion and Society at the University of Victoria was co-sponsored by the Centre for Global Studies, the Centre for Forest Biology, the Institute for Integrated Energy Systems, and the Climate Modelling Group, all at the University of Victoria. Special thanks are due to Connie Carter, former administrator, Centre for Studies in Religion and Society at the University of Victoria, for preparing the manuscript for publication.

—Harold Coward

Introduction

Jan Zwicky

WEATHER. WHEN I WAS A KID on the farm in Alberta, everything depended on it: what you put on in the morning and what you ate for breakfast; what you did at any given hour of the day and whom you did it with; my grandfather's, my mother's, and my own mood. And we talked about it constantly—"Did you notice? Clouding over in the west," "If it clears off now, she'll freeze," "It's already over 70° (in those days we still used Fahrenheit), gonna be a scorcher"—the hourly, sometimes minute-by-minute, warp on which our lives were strung. Much later in my life, my job often had me on the road late at night between Saint John and Fredericton. To keep myself awake, I listened to the radio. My favourite part was always the late-night marine forecast: although I'd never been in a fishing boat in my life, nor seen the straits, banks, and fans the reports named, there, alone in the dark on a New Brunswick highway, I felt profoundly at home—connected, through the significance of weather, to a life like my own.

Before the move to the Maritimes, I'd spent several years in southwestern Ontario, a region renowned for its muggy summer heat and slow-witted thunderstorms. One July afternoon in 1988, I was hanging out a load of laundry. It had been hot for days, temperature parked in the thirties, but around noon a breeze had sprung up and I figured we were in for a change. It was, in fact, downright windy by the time I got the clothes to the line—but, to my surprise, there wasn't a cloud in sight. There was a haze on the horizon, but I realized it was dust. As I pegged, struggling with the shirts and towels as they jerked in the wind, I was overcome with unease.

It was hot, it was incredibly hot. I paused, and looked again for clouds on the horizon. Nothing. And then it struck me: the wind *itself* was hot. It was actually making things hotter. I touched the clothes I'd hung out first—already dry. And the phrase entered my mind: "it's *too* hot." I didn't know exactly what I meant by "too," except that the hair on the back of my neck was rising. This wind was like none I had experienced. It didn't belong. By the time I got back into the house, I was shaking. I had to send my partner out to get the clothes in off the line.

Of course, I told myself that evening, one hot day does not a global catastrophe make. What I'd experienced that afternoon was at the upper end of the range for Ontario, yes, but not completely outside it. It was proof of nothing. Still, there had been something about the character of that wind: its dryness, its persistence, the absence of clouds, the way it had come, by itself, out of the southeast. I still can't find the words to capture exactly what struck me as wrong. But I can say that the problem wasn't just the temperature; it was some complex cumulative effect at the peripheries of awareness, the touch of something weird.

I've caught the same faint whiff of horror a number of times in the last few years. Two years ago, my mother rang from the farm in Alberta. "Remember I said I was moving the outhouse? Herb was in to dig the hole today." Pause. "It was dry six feet down. Like dust." That's all she said. I couldn't think of anything to say either. She knew I knew that my grandfather had maintained they'd made it through the drought in the 1930s because there was always a touch of moisture, from somewhere, in the osmotic lacustrine clays that underlay the gumbo he farmed. And last summer, I sat in the porch of the new house, watching the topsoil from the garden lift and blossom in the wind like smoke. I'd read *As for Me and My House*. I knew it had been as bad, or worse, in Saskatchewan, and they'd survived. But this wasn't happening in Saskatchewan. It was happening to the north and west, on what was supposed to be the cold, damp boreal fringe of the Great Plains, not its dry, sunny heart. Sitting there, watching the soil billow and drift, I felt a sick grief. I'd have done almost anything to make it stop; and there was nothing I could do.

✵

But let's return a moment to my earlier observation that one hot day in southwestern Ontario is not tantamount to proof that global warming is on our doorsteps. Doesn't the same thing apply to the Prairies in spades? It's always been touch-and-go there. The archaeological record suggests that the region was subject to prolonged and severe drought during the

Hypsithermal—the peak of the warming trend that followed the Wisconsin glaciation (E.G. Walker, 1988). And come to that, what *about* the Ice Age? Isn't the current interglacial supposed to be coming to an end (Pielou, 1991, esp. p. 12–15)? (Maybe a little human-induced warming wouldn't be such a bad thing!) In certain regions—like the Palliser triangle—it looks like the "anomalous" weather was the wet stuff that suggested to John Macoun in 1879 that European-style agricultural settlement could be sustained there, not the dry stuff that's since made it difficult (Herriot, 2000, pp. 154–55).[1] So, yes, maybe there's a drought underway, but 1) there's no telling how bad it will be, and 2) 'twas ever thus. A person might be upset about bad times down on the farm, but intuitive alarm bells of the sort I've described are not grounds for thinking the climate is changing. They are, after all, merely intuitions. They're not science.

So what does science say about climate change and the sorts of worries I've been having? Here's an analogy. Think of the way a pointillist painting works. At a distance of ten feet, even six, hand appreciatively on jaw, it all makes sense: we can see the tree, the sailboat, the river—the larger shapes and patterns that determine the meaning of the painting. But now imagine you are an ant crawling on the painting's surface. You find yourself on a splotch of beige. Does this mean you are on the pavements of the promenade? Or perhaps the walls of the clock tower? By no means. You could be on a highlight on some passerby's black hat. No single dab of colour is impossible anywhere; and sometimes the overall effect is the result of a surprising series of local juxtapositions. Individual dots of colour have nothing to do with what we perceive when we stand back. What science tells us is that weather is to climate what the dots are to the big picture.

The main consequence of the analogy cannot be emphasized too strongly, and bears repeating: the ant is not justified in drawing climatological conclusions from any given splotch of paint it finds itself on. Even if it finds itself in an area with a lot of beige dots, it might be on a bit of cloud showing through the branches of a tree, or a patch of sunlight reflecting off the water. If it's a very big painting—and climate is a *very* big painting—it may take generations of ants, and pooled data from ants in other regions, to begin to firm up one's guess. Climate, although its foundation is the weather that surrounds our daily lives, is actually something quite abstract—a collection of aggregated averages that have no concrete instantiation. Climate is like the 2.3 children possessed by the average Canadian family; weather is the actual kids (integers only, please) fooling around in the backyard.

A second consequence of the analogy may also be worth drawing out, if only because we may lose sight of it in insisting on the first. It's not as if there's *no* connection between climate and individual weather events. If the ant finds itself in an area with a lot of beige dots, and then suddenly crosses into an area with mostly blue ones, it would be reasonable for it to speculate that it had crossed a boundary of some sort and had entered a different region of the painting, even if that difference is merely the difference between water with sunlight glare and water without. That is, although the connection between climate and local weather is not immediate, it's there. Just as the ants never experience the big picture, we never directly perceive climate; but our experience of weather, especially over the long term and in connection with other things like native vegetation, is nonetheless relevant to our sense of what the larger pattern might be.

Moreover, despite trends in the style of Western European thinking over the last 500 years, we remain a species that tends to make sense of immediate experience by picking up on patterns that inform it. We share this tendency, I believe, with every other species on the planet, and, although our interest in developing our capacities for systematic reconstruction has blunted our trust in gestalt perception, the ability to pick up on big pictures (or changes in big pictures) from 'mere details' remains part of our genetic inheritance. In terms of the analogy: so we're plodding merrily along, first a beige splotch, then a blue one, then a beige, then a blue, ah yes, we're out on the river amid ripples scintillating in sunlight, then a red one, then a beige, then a—what? red?... hmmm—then a blue, then another red, I guess it must be sunset... Wait a minute. What if I'm off course? Aren't those blue-and-beige alterations a bit too *regular* for ripples? Uh-oh: that red's not red, it's rust! I'm on the *tent awning* way on the other side of the river and those blotches of colour are *birds*.

Is intuition of this sort infallible? Of course not. Is its fallibility a reason never to pay attention to what it's suggesting? No. Intuition told me it was "too" hot, that July afternoon in Southwestern Ontario, and it tells me now it's "too" dry on the farm. As with any gestalt perception, the reasons for these assertions are hard to articulate, and I'm deeply conscious that as intuitions they won't, as it were, stand up in court. The net effect is that I've got a lot of questions. My attentiveness to weather patterns has been heightened. I've stopped plodding merrily along. I'm scanning the horizon. I'm listening. The essays in this book lay out some of the basics I've been hoping to learn.

✲

Before we turn to a summary of their contents, let me offer yet another analogy that highlights a related, but more extended, sequence of points: to be told that global warming is a reality is the cultural equivalent of being diagnosed with a life-threatening disease. The first point is that even if you do have, say, cancer, this doesn't mean that every cough or twinge is a symptom—though initially, it's natural to be jumpy and to see bad news everywhere. The profile of a given disease, like the profile of health, is an aggregate of common and not-so-common features—an "ideal" case that is rarely, if ever, instantiated. Often a given disease is accompanied by a high fever—but you may not get one. It rarely produces hair loss—yet you may go bald. Even if the diagnosis is confirmed beyond the shadow of a doubt, we can at best say only what is likely to occur in any given case. And the same thing is true of global warming were we to try to say what might happen in any particular region over any given stretch of time.

The second point is that certain things—a series of nasty colds, severe digestive upsets, shortness of breath—*are* reasons for concern. Further, if you have a life-threatening disease, you can indeed expect to experience some bad stuff, especially in the later stages, if you leave the disease untreated.

The third point—connecting to my earlier observations about gestalt twitchiness—is that it is not at all uncommon for people to become concerned well before the onset of major symptoms or for apparently trivial reasons. Carlo Ginzburg, in a discussion of scientific method (1980), has linked such diagnostic sensitivity both to connoisseurship and to "detective" ability, that is, to a general sensitivity to the "telling" detail. The existence of a genuine capacity of this sort need not, in terms of our analogy, be taken as licensing snake oil salesmen; rather, often what a sensitive diagnostician can pick up is simply that something's not quite right, that more research is required. It can be the goad to the labour of detailed empirical investigation that otherwise might not be deemed necessary.

The analogy is also useful in underscoring that the diagnosis of global warming does not necessarily mean the patient is going to die. The operative word here is life-*threatening*: timely intervention may result in a cure, or at least a condition that can be "managed." However, spontaneous remission in the absence of intervention is very rare. The sort of diagnosis we're talking about is a serious matter and in the face of it, "business as usual" is not a serious option. Minimally, some sort of treatment—in the Western tradition, a course of medicine, say—is required. Maximally, a person may be persuaded to undergo invasive surgery, accept ongoing pharmacological support, and undertake a complete lifestyle overhaul in order to survive. It is not surprising that, in such circumstances, interest in non-Western therapeutic approaches and traditions is often intense.

Of course, it's also possible for interest in any sort of therapy to be entirely absent. To pursue treatment, one must first be willing to acknowledge one has a disease. Denial, as we all know, is a common phenomenon and may be manifest both before and after diagnosis. Before diagnosis, it consists in a refusal to go to the doctor to find out if anything is wrong: *especially* if we've been heavy smokers for years, we may be unable to take the persistent cough and shortness of breath seriously. It *has* to be a stubborn cold; the other thought is just too scary. After diagnosis, denial can take the form of a rejection of test results ("Everybody knows you get a lot of false positives, eh?") or a downplaying of the seriousness of the consequences ("Hey, it's nothing—a quick trip to the chop shop and I'll be right as rain" or "They can cure anything these days").

But the fundamental question in all this is surely *Is global warming actually occurring? Have we got a confirmed diagnosis?* The unequivocal answer laid out by Andrew Weaver in chapter 2 of this book is "yes." The scientific evidence is overwhelming that climate change is underway. Weaver also shows that the data now available provide an unequivocal answer to the question of what is causing this change: anthropogenic greenhouse gas emissions, or, in slightly cruder terms, things that industrialized human beings are pumping into the atmosphere.

So that's the bad news. The next question a person usually asks once they're over the initial shock is *Does this mean I'm going to die?* Here the news is a little better. Many projections suggest the case is not necessarily terminal. The factors are immensely complex, but there is reason to hope global warming's worst ravages might be avoided if timely treatment is undertaken.

The remaining chapters in Part I take up two other common questions: *Is there usually a lot of pain?* and: *What about my specific case? How bad's it going to be?* In chapter 3, Steve Lonergan documents the occurring and expected consequences of climate change for human populations globally, emphasizing that those who can least afford to protect themselves are likely to be hardest hit. In the fourth chapter, Jim Bruce and Stewart Cohen narrow the focus to Canada, providing regional data on changes that have already occurred and offering projections for the future.

OK, then, doc—what do we do? Anything, just so long as I can live! The chapters in Part II of the book are an attempt to provide an overview, from a Western perspective, of treatment options. Here, the news is also mixed, and the factors complex. The best thing for the patient (a.k.a. the world's industrialized and industrializing nations) would be to lose a lot of weight and quit smoking. But it's not clear the patient is willing to do what's best. *Hey, doc, wait a minute—isn't there some other way? What if I switch to filter? My brother-in-law heard they figured out how to make cigarettes without tobacco—maybe I could smoke them?* The questions may sound silly, but the point is that to a culture that is severely addicted, the cure can sound worse than the disease. Not only are most Western governments and large industries hesitant about the prospect of a major economic overhaul, most individuals are not (yet!) willing to give up on many of the goods and services that greenhouse-gas-emitting technologies provide. Nor do industrialized and industrializing nations seem poised to embrace sharply negative population growth, which would be a means of tackling the problem from the other end. These, then, are the constraints on the possible "fixes" explored in Part II.

Increasing earth systems' abilities to absorb carbon—adding to our "carbon sinks" with forestation projects, for example—can help, according to Nigel Livingston and G. Cornelis van Kooten in chapter 5, but only in the short run. Ged McLean and Murray Love, in chapter 6, argue that alternative energy sources for the goods and services we want are indeed available—but not as handily, or immediately, or as free of nasty consequences as we might wish. In chapter 7, G. Cornelis van Kooten takes up the question of the purely economic costs of slowing or reversing carbon dioxide emissions. Cost-benefit analysis rests on assumptions that are contested in some quarters, but it indicates that the costs of significant reductions in greenhouse gas emissions are greater than industry and government will be willing to pay. As a result, van Kooten suggests we concentrate our attention on attempts to adapt to climate change rather than on attempts to mitigate it. In chapter 8, Stewart Cohen and company offer another reason to concentrate on adaptation when they return to one of the bits of bad news from chapter 2: there's a lag time between when the greenhouse gases enter the atmosphere, and when their effects are felt, so that even if we were to quit cold turkey *today*, we would still need to cope *in the future* with what we've already done. Regardless of policy decisions on mitigating climate change, then, adaptation strategies are going to be required, and this raises the subsidiary question of how prepared we can financially afford to be. In the final chapter of Part II, Alastair Lucas discusses the legal challenges of designing and implementing policy to deal with climate change.

Part III turns from the question of what it's possible to do to the question of what ought to be done. In chapter 10, Jim Bruce and Doug Russell survey the challenges of formulating public policy at a number of different levels of government in Canada. In chapter 11, the history and effectiveness of the Kyoto Protocol, a multilateral agreement to reduce global greenhouse gas emissions, is explored by Gordon Smith and David Victor. Their conclusion is that Kyoto as it stands will not do the job we need it to do, and that Canada ought therefore to recommend and implement significant structural changes. In the final chapter of the book, Harold Coward looks at responses to global climate change from the point of view of our individual ethical orientation to the earth as a whole. He discusses a range of secular and spiritual paradigms, each of which encapsulates a set of fundamental moral values that can be used to orient our attitudes and ground individual choices.

※

Climate change is only one of several human-induced environmental—what?... crises? difficulties? challenges? Any noun I might choose has political spin, defines allegiances, presupposes a point of view. Robert Bringhurst puts the point succinctly:

> *Being will be here.*
> *Beauty will be here.*
> *But this beauty that visits us now will be gone.*
> —(1995, p. 200)

It is hubris to imagine our species can destroy everything, or even everything that matters to it, just as it is hubris to imagine we are what evolution is "for," or that human interests are distinct from and ontologically superior to all others.

"Why is there something rather than nothing?"—A question that has no answer, but one that is rooted in a fact that has absorbed and moved great thinkers from Lao Tzu to Heidegger. Which is not to say that you have to be a philosophical genius to experience astonishment that things *exist*: it's a common experience among the naturalists and poets of my acquaintance. And I'll bet dollars to doughnuts many of the folks leaning in, listening to those late-night marine forecasts, felt it from time to time. Our astonishment is the mark of our mortality. Is-ness *is*, always; but *what* is, *this*, is here only now. The love we feel for concrete particulars—a stand of birch, a stretch of river, no less than other human beings—is as biologically basic as our sexual mode of reproduction. We must love what dies and we must love *because* we die. Plato, like other religious thinkers in other traditions, sought to ease the pain attendant on this inheritance by encouraging us to fix our erotic gaze on eternity, on the non-particularized being that informs everything that is. But me, I'm with Herakleitos: "The things of which there is seeing, hearing, and perception, these do I prefer" (Diels, 1934, Fr. 55). I would be the last to deny the power of universal, atemporal being; it's just that because I'm human—that is, because I love and die—it's only half the story. "Nameless:" says the *Tao Te Ching*, "the origin of heaven and earth./ Naming: the mother of ten thousand things" (1993, chap. 1). Those ten thousand things are the other half of the story. They are the manifestations through which the mystery flows, without which it would be invisible, of which we are one. We hope because, quite apart from the philosophers, we have good reason to believe that beauty will be here: there will be trees and grass and rivers and, unless we are staggeringly stupid, a few humans around to appreciate them. We grieve because we also have reason to believe that this beauty—at least some among these copses, these grasslands, these shorelines—will not survive. That is what this book is about: the grounds for that hope, and that grief.

Notes

1 Herriot compares reports of three early European visitors to the region, noting differences in rhetorical style which may also have influenced the Canadian government's decision to promote European colonization. The three accounts in question are to be found in Palliser (1859), Hind (1971), and Macoun (1882).

References

Bringhurst, Robert. (1995). New World Suite N° 3. *The Calling: Selected Poems 1970–1995*. Toronto: McClelland & Stewart.

Diels, H. (1934). *Die Fragmente des Vorsokratiker.* 5th ed. Rev. W. Kranz. Berlin: Weidmann.

Ginzburg, Carlo. (1980). Morelli, Freud and Sherlock Holmes: Clues and Scientific Method. Trans. Anna Davin. *History Workshop Journal* 9: 5–36.

Herriot, Trevor. (2000). *River in a Dry Land: A Prairie Passage.* Toronto: Stoddart.

Hind, Henry Youle. (1971). *Narrative of the Canadian Red River Exploring Expedition of 1857 and of the Assiniboine and Saskatchewan Exploring Expedition of 1858.* Vol. 1. Edmonton: Hurtig.

Macoun, John. (1882). *Manitoba and the Great North-West.* Guelph: World Publishing.

Palliser, John. (1859). Papers Relative to the Expedition by Captain Palliser of That Portion of British North America Which Lies between the Northern Branch of the River Saskatchewan and the Frontier of the United States; and between the Red River and Rocky Mountains. Paper presented to both Houses of Parliament by Command of Her Majesty, June 1859. London: G.E. Eyre and W. Spottiswoode.

Pielou, E.C. (1991). *After the Ice Age: The Return of Life to Glaciated North America.* Chicago: University of Chicago Press.

Tao Te Ching. (1993). Trans. Stephen Addiss and Stanley Lombardo. Indianapolis: Hackett.

Walker, Ernest G. (1988). The Gowen Site: A Mummy Cave Occupation within the City Limits of Saskatoon. In *Out of the Past: Sites, Digs and Artifacts in the Saskatoon Area,* ed. U. Linnamae and T.E.H. Jones, 65–74. Saskatoon: Saskatoon Archaeological Society.

Part I

What's [Going to] Happen[ing]?

The Science of Climate Change

Andrew J. Weaver

Introduction

IN JANUARY 2001, the United Nations Intergovernmental Panel on Climate Change (IPCC) released a report stating that there is now new and stronger evidence that most of the climate warming observed over the last 50 years is attributable to human activities. This powerful statement by the world's leading climate scientists sends a strong signal to governments that informed policy is urgently needed to determine a course of action for the future. To set this target, researchers must reduce uncertainty in climate projections and quantify the socio-economic impacts of climate change. They must also develop the policies and mitigation technologies that will most effectively achieve the appropriate levels of net greenhouse gas emissions and develop the adaptation strategies that will respond to the consequences resulting from those choices.

This chapter presents a review of the science of climate change, starting with a discussion of the 200-year history of the science leading up to our present-day understanding of global warming. Since much of the observational evidence and many future projections of climate change are derived from the IPCC Third Assessment Report, a brief review of the history behind the formation of the IPCC is given before this chapter turns to climate change detection and attribution: the search for an anthropogenic (human-induced) warming signature above a background of natural variability. A summary is given in section 7 and concluding remarks offered in section 8.

History of the Science of Global Warming

A common misconception is that the link between increasing levels of atmospheric carbon dioxide (CO_2) and global warming has only recently been realized. In fact, Dr. James Hansen of NASA's Goddard Institute for Space Studies is often credited in the media as being the "father of climate change theory."[1] This labelling has apparently occurred in response to Hansen's being called before the US Senate Committee on Energy and Natural resources to testify on 23 June 1988. At that time, he argued that he was 99% confident that the greenhouse effect had been detected and that it was changing our climate. This now famous testimony occurred nearly 100 years after the link was drawn between CO_2 and the earth's temperature (see Christianson, 1999, for a historical review). It was perhaps one of the first times the issue directly entered the US political arena and so received a high profile in the media.

In fact, in 1824, Jean-Baptiste-Joseph Fourier (a well-known French mathematician) hypothesized that the atmosphere blocks outgoing radiation from the Earth and reradiates a portion of it back, thereby warming the planet (Fourier, 1824). Swedish Nobel laureate Svante Arrhenius drew upon Fourier's work, as well as that of American astronomer Samuel Langley and Irish polymath John Tyndall, to develop the first theoretical model of how atmospheric CO_2 affects the Earth's temperature (Arrhenius, 1896). In 1938, the British coal engineer George Callendar argued that since 1880 the Earth had warmed by about 1°F, and he predicted that this would double in the next half century.

In a seminal paper, Revelle and Suess (1957) argued that the oceans could not absorb the human emissions of CO_2 as fast as they were being produced. Revelle and Suess further noted that this would leave the human-released CO_2 in the atmosphere for centuries and stated, "Human beings are now carrying out a large-scale geophysical experiment of a kind that could not have happened in the past nor be reproduced in the future." They further argued that "we are returning to the atmosphere and oceans the concentrated organic carbon stored in the sedimentary rocks over hundreds of millions of years."

The first sophisticated atmospheric modelling studies aimed at investigating the climatic consequences of increasing atmospheric CO_2 were conducted at the NOAA Geophysical Fluid Dynamics Laboratory, now in Princeton, New Jersey. Imbedded within the abstract of a paper written by Manabe and Weatherald (1967) was the conclusion, "According to our estimate, a doubling of the CO_2 content in the atmosphere has the effect of raising the temperature of the atmosphere (whose relative humidity is fixed) by about 2°C." This early work yielded a projection consistent with the IPCC's 1996 "best guess" estimate of 2°C warming by 2100 (where atmospheric CO_2 is projected to double, relative to preindustrial levels, by year 2070) (IPCC, 1996).

By the early 1980s, the issue of climate change began to move from the scientific to policy agendas. Several scientific assessments of the relationship between CO_2 and climate began to appear.[2] On the international scene, it is apparent that a series of conferences and reports organized by the United Nations Environment Programme (UNEP), International Council of Scientific Unions (ICSU), and the World Meteorological Organization (WMO) were especially influential. The Second Joint UNEP/ICSU/WMO International Assessment of the Role of Carbon Dioxide and Other Greenhouse Gases in Climate Variations and Associated Impact, which took place in October 1985 in Villach, Austria, was particularly important in this regard.

In summary, the theory of global warming is based on elementary principles of physics—principles that have been around for more than a century: warm climates can't be maintained unless there are greenhouse gases to block outgoing radiation; cold climates can't be maintained unless there is a depletion of greenhouse gases (see fig. 2.1). If the levels of these gases are changed, the result is a radiative forcing (see next section) to which the Earth system must respond. Global warming is not a new issue that appeared in 1988 when James Hansen gave testimony before the US Senate, but rather, it is deeply rooted in two centuries of science. National and international assessments have been conducted on the topic since the early 1980s. Most recently, this task has been charged to the United Nations IPCC.

The Intergovernmental Panel on Climate Change Process

The Intergovernmental Panel on Climate Change was established in 1988 by the World Meteorological Organization and the United Nations Environment Programme as a means to assess the potential problem of global climate change. It is a UN organization governed by UN regulations with a mandate most recently reaffirmed in Vienna in October 1998:

> The role of the IPCC is to assess on a comprehensive, objective, open and transparent basis the scientific, technical and socio-economic information relevant to understanding the scientific basis of risk of human-induced climate change, its potential impacts and options for adaptation and mitigation. IPCC reports should be neutral with respect

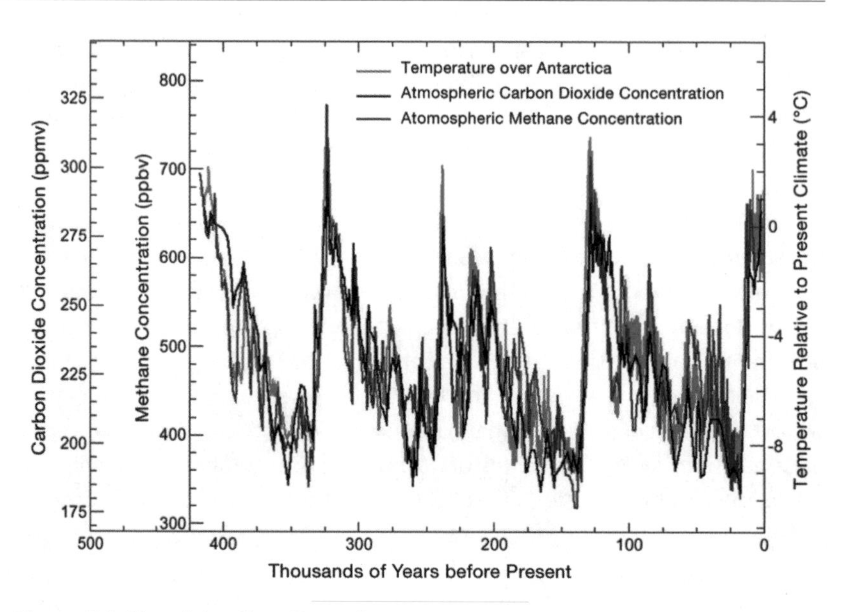

Figure 2.1. Vostok Ice Core Record

Variations in atmospheric temperature derived from isotopic data and concentrations of atmospheric carbon dioxide (black) and methane (grey) from Vostok, Antarctica ice core records. Notice that the current level of atmospheric CO_2 (368 ppm) is >20% larger than at any time during the last 400,000 years. Similarly, current levels of atmospheric methane CH_4 (1750 ppb—too far off the scale to be plotted) are more than double the maximum value found in the 400,000-year record. The increase in CO_2 from 280 ppm to 368 ppm over the last 150 years, primarily due to fossil fuel burning, is about the same as the increase from the depths of the last ice age (21,000 years ago) to 1750 (190 ppm to 280 ppm). *Source*: IPCC, 2001.

> to policy, although they may need to deal objectively with scientific, technical and socio-economic factors relevant to the application of particular policies. (IPCC, 1998)

To address this mandate, the IPCC oversees three working groups (WGI, WGII, WGIII) aimed at assessing the science, socio-economic impacts and adaptation, and mitigation aspects of climate:

- WGI: assesses the scientific aspects of the climate system and climate change.
- WGII: addresses the vulnerability of socio-economic and natural systems to climate change, negative and positive consequences of climate change, and options for adapting to it.
- WGIII: assesses options for limiting greenhouse gas emissions and otherwise mitigating climate change.

A common public misconception is that the IPCC working groups undertake their own independent research. This is not the case: they only provide an assessment of the peer reviewed literature, although they make reference to published technical reports. IPCC does not consider Web sites, or newspaper opinion pieces or editorials to have passed the standards set by the peer-review system, and so will not include these in the assessments.

There have now been three formal IPCC Assessments of Climate Change. The first, in 1990, led to the creation of the Intergovernmental Negotiating Committee for a UN Framework Convention on Climate Change by the UN General Assembly. The second assessment, in 1996, was formally used in the negotiations leading up to the adoption of the Kyoto Protocol to the UN Framework Convention on Climate Change at the Third Conference of Parties in 1997. The Kyoto Protocol requires Canadian greenhouse gas emissions to be 6% below 1990 levels in the period spanning 2008–2012. The third IPCC assessment was completed in 2001, and the fourth assessment will be completed in 2007.

While the IPCC assessments ultimately enter the political arena, the actual writing of the assessment itself is free from political interference. In the third assessment, for example, 120 of the world's leading climate scientists wrote the WGI document, with contributions from over 500 other climate scientists. The content of an individual chapter was chosen exclusively by the lead authors of that chapter, in consultation with the lead authors of other chapters (to ensure that there was no duplication). The final report underwent review three times by more than 300 experts in the field. This review process included an informal review by all lead authors; a review; by experts in the field; an additional expert review and government review. The third draft of the document was put together after the IPCC meeting in Victoria, British Columbia, and was sent to United Nations member states for approval in Shanghai in January 2001. Final changes were made to the "Summary for Policy-makers" in Shanghai as a consequence of feedback from UN member states.

As noted above, the formal charge of WGI is the assessment of available information on the science of climate change and on its association with human activities. More specifically,

> In performing its assessments WGI is concerned with: developments in the scientific understanding of past and present climate, of climate variability, of climate predictability and of climate change including feedbacks from climate impacts; progress in the modelling and projection of global and regional climate and sea level

> change; observations of climate, including past climates, and assessment of trends and anomalies; gaps and uncertainties in current knowledge. (IPCC—WGI, n.d)

In what follows, I will draw heavily from the assessment that arose from this IPCC WGI process. In particular, I will focus on the key findings of chapter 2, "Observed climate variability and change," in the next section. Then I will highlight the most important aspects of chapter 9, "Projections of future climate change," and chapter 12, "Detection of climate change and attribution of causes," respectively.

Observational Evidence of Climate Change

Radiative forcing of climate

The Earth is said to be in a global radiative equilibrium if the total amount of energy received from the sun, averaged over a few decades, equals the total amount of energy emitted by the earth to space. A change in the average net (incoming minus outgoing) radiation at the top of the atmosphere is defined as a *radiative forcing*. A positive radiative forcing acts to warm the Earth's surface, while a negative radiative forcing acts to cool it. That is, a radiative forcing is a disturbance in the balance between incoming and outgoing radiation, and, over time, the climate system (fig. 2.2) responds to try and re-establish global radiative equilibrium.

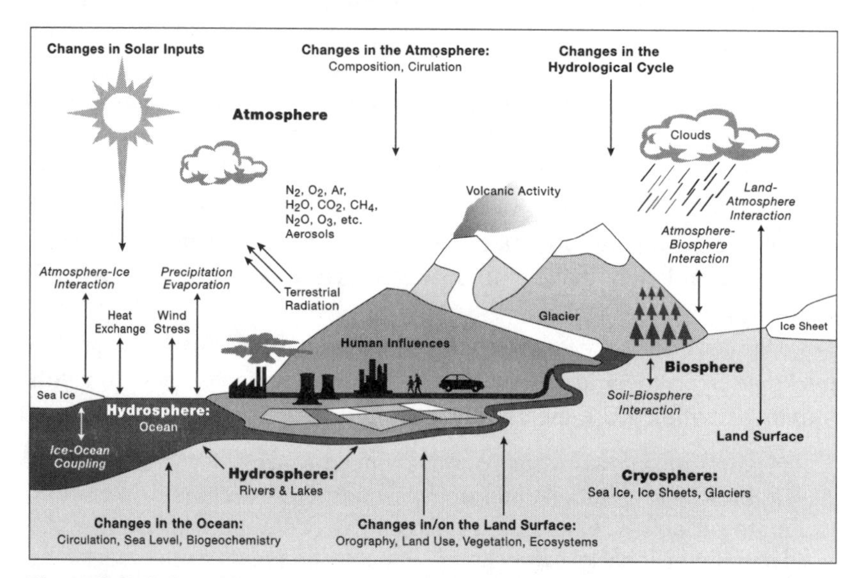

Figure 2.2. Schematic representation of the climate system *Source*: IPCC, 2001.

Carbon dioxide (CO_2), methane (CH_4), and nitrous oxide (N_2O) are examples of greenhouse gases whose increase over the last 150 years has created a positive radiative forcing (fig. 2.3). Aerosols, which are tiny liquid or solid particles in the atmosphere, are most often considered to provide a negative radiative forcing (e.g., sulphate aerosols released in the combustion of coal). These and other aerosols affect the radiation balance of the Earth by both directly scattering incoming radiation back to space and indirectly affecting the formation, lifetime, and properties of clouds.

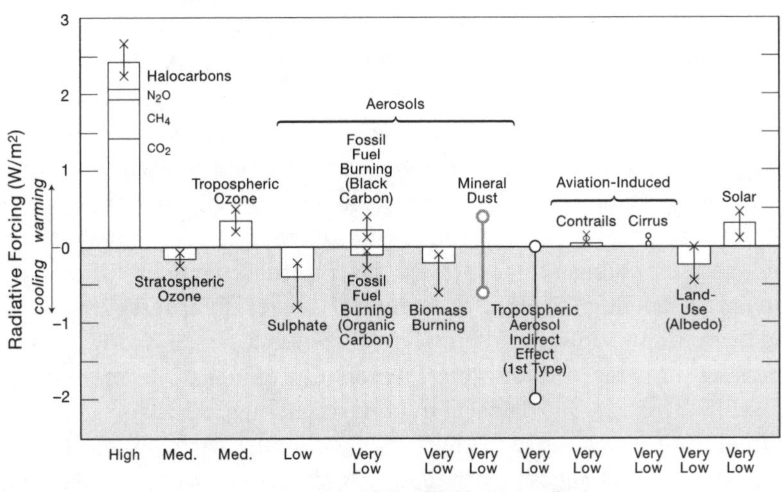

Figure 2.3. Global and Annual-Mean Radiative Forcing (W/m²) for Various Agents from Pre-industrial Times (1750) to the Present (late 1990s)

The height of the rectangular bar denotes a best-estimate value, while its absence denotes that no best estimate is possible. The vertical line about the rectangular bar with "×" delimiters indicates an estimate of the uncertainty range, for the most part guided by the spread in the published values of the forcing. A vertical line without a rectangular bar and with "0" delimiters denotes a forcing for which no central estimate can be given owing to large uncertainties. A "Level of Scientific Understanding" index is accorded to each forcing, with high, medium, low, and very low levels. This represents the subjective judgment about the reliability of the forcing estimate. The well-mixed greenhouse gases (e.g., carbon dioxide, methane, and nitrous oxide) are grouped together into a single rectangular bar. The sign of the effects due to mineral dust is itself uncertain. The indirect forcing due to tropospheric aerosols as well as the forcing due to aviation via their effects on contrails and cirrus clouds, is poorly understood. The forcing associated with stratospheric aerosols from volcanic eruptions is highly variable over the period and is not considered for this plot. It is emphasized that the positive and negative global-mean forcings cannot be added up and viewed a priori as providing offsets in terms of the complete global climate impact. *Source*: IPCC, 2001.

Of course, there are significant differences in the atmospheric residence time of individual greenhouse gases and aerosols due to natural removal mechanisms. Tropospheric aerosols, for example, stay in the atmosphere only a few days because they are effectively scavenged by precipitation. Stratospheric aerosols, such as those released during volcanic eruptions, have a residence time of up to a few years since they must first descend, through gravity, into the troposphere before they can be scavenged by precipitation. CO_2 has a lifetime between 50 and 200 years, methane about 12 years, nitrous oxide about 120 years, CFC-11 about 50 years and a perfluorcarbon (another greenhouse gas) about 50,000 years.

Finally, it is important to note that the Earth system does not instantly reach radiative equilibrium once a radiative forcing is applied. The slow time scales inherent in the system, such as those associated with the ocean, lead to a lag of several centuries before quasi-equilibrium can be reached. As a specific example, Wigley (1988) considered the climatic effects of the ratification of the Kyoto Protocol. He showed that if all countries followed their baseline changes after 2010 (i.e., all countries met their Kyoto targets but did no more for the rest of this century), the resulting best guess warming of 2.08°C by 2100 would only be reduced to 2.0°C. Similarly, the best guess sea-level rise of about 50 cm would only reduce to 48 cm.

Unlike the case for glacial to interglacial changes which occur on millennial time scales (fig. 2.1), thereby allowing the Earth system time to equilibrate with changes in the radiative forcing, the current rate of change in radiative forcing is very rapid. As such, there is inevitable warming in store as the earth system attempts to equilibrate with the higher levels of greenhouse gases. In terms of global warming, therefore, the real policy question that needs to be addressed is what we as a society consider to be an acceptable level of future warming (see chap. 3).

Observed changes in surface air temperature

Several researchers around the world have independently put together global data sets of surface air temperatures from the instrumental record. All of these researchers have either used only non-urban locations or corrected the urban data for the urban heat island effect (temperatures within cities will naturally warm as the cities grow).

The globally averaged surface air temperature has increased by about 0.6 ± 0.2°C over the twentieth century. Most of this warming has occurred during two periods: 1910–1945 and 1976–2000 (fig. 2.4a, 2.5). Very recently, proxy data from, for example, boreholes, corals, and tree rings have allowed the reconstruction of northern hemisphere temperatures back as far as

1000 AD. Of particular importance is the fact that reconstructed and instrumental records agree remarkably over their common period. In the last 1,000 years, the twentieth century is the warmest century and the 1990s the warmest decade. The top 10 warmest years in descending order are 1998, 2003, 2002, 2001, 1997, 1995, 1990, 1999, 2000, and 1991. A recent US National Academy of Sciences report (NRC, 2000), has reaffirmed the validity of the global surface temperature record, and new published scientific studies (e.g., Prabhakara et al., 2000; Christy et al., 2001) are beginning to show similar warming trends in satellite and surface-based data sets (0.13°C per decade).

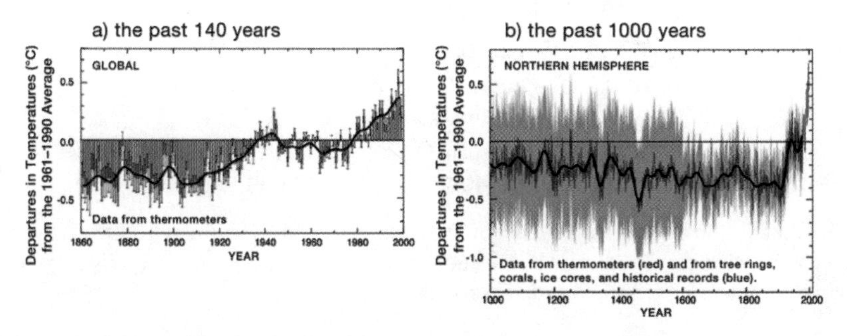

Figure 2.4. Variations in the Earth's Surface Temperature, Globally and for the Northern Hemisphere:

a) the combined 1861–1999 annual mean land surface air and sea surface temperature anomalies relative to the 1961–1990 mean: the black curve gives a decadal average; b) reconstructed northern hemisphere surface air temperature changes since 1000 AD (blue). The black curve represents a 40-year smoothed version of the blue curve and the red curve comes from (a). Standard error limits are shown by bars in (a) and by gray shading in (b). *Source*: IPCC, 2001.

A common misconception is that global warming implies warming everywhere by about the same amount. This is not the case and there are, in fact, regions where the earth has cooled over the twentieth century (fig. 2.5a). Warming on the global scale is either amplified or reduced through local feedbacks. In general, the warming is much greater over land compared to oceans as the oceans have a higher heat capacity and can sequester heat to great depths (see, for example fig. 2.5d). Warming is also generally greater at high latitudes than at low latitudes, due to the existence of a powerful positive feedback involving the albedo[3] of snow and ice. That is, as snow and ice cover retreat, as has been observed over the twentieth century (especially since 1979), the land surface darkens and so does not reflect as much radiation back to space. In the case of sea ice, the observed reduction in areal extent also exposes more of the ocean to the

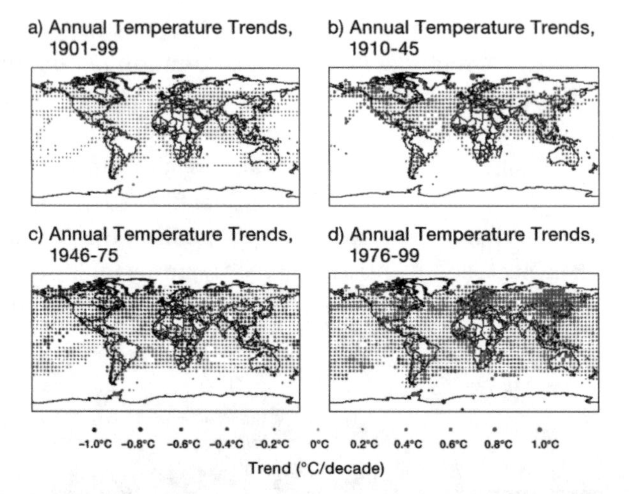

Figure 2.5. Annual Mean Temperature Trends (°C/decade) for the Periods: a) 1901–1999; b) 1910–1945; c) 1946–1975; d) 1976–1999

The magnitude of the trend is given by the area of the circle and the sign of the trend is positive (warming) if the circle is red, and negative (cooling) if the circle is blue. A green circle indicates little or no trend. *Source*: IPCC, 2001.

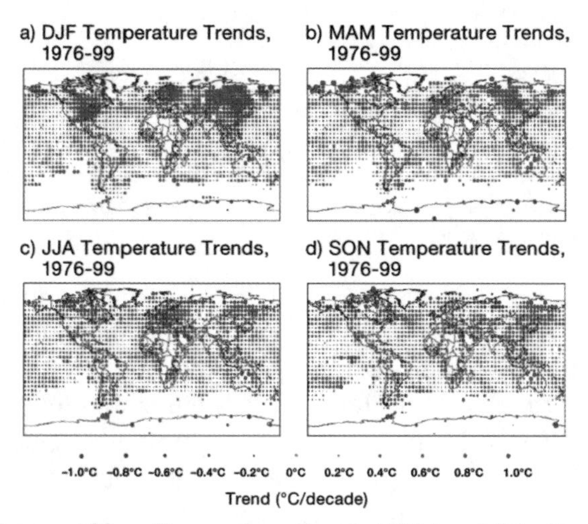

Figure 2.6. Seasonal Mean Temperature Trends (°C/decade) for the Period 1976–1999

a) Winter: December, January, February; b) Spring: March, April, May; c) Summer: June, July, August; d) Autumn: September, October, November. The magnitude of the trend is given by the area of the circle and the sign of the trend is positive (warming) if the circle is red and negative (cooling) if the circle is blue. A green circle indicates little or no trend. *Source*: IPCC, 2001.

atmosphere, thereby allowing the warming of the atmosphere through heat loss from the ocean.

The warming is also amplified in the winter, and to a lesser extent in the spring, over land since the snow albedo effect is greatest at this time of year (fig. 2.6). In addition, the warming trend over land since 1950, on average, has been about twice as fast at night compared to the day (fig. 2.7). That is, nighttime low temperatures have increased twice as fast as daytime high temperatures (0.2°C per decade versus 0.1°C per decade). It is currently thought that this decrease in diurnal temperature range is associated with the observed increase in cloud coverage since 1950. Clouds act to dampen diurnal temperature variations by reflecting incoming solar radiation during the day, and absorbing and re-radiating outgoing longwave radiation from the earth at night. Similarly, the direct effect of tropospheric aerosols only acts in the day (when there is incoming solar radiation to backscatter) while greenhouse gases are effective both day and night. As with the warming trends, there are some regions where in fact the diurnal temperature range has increased since 1950, although most regions show a reduction.

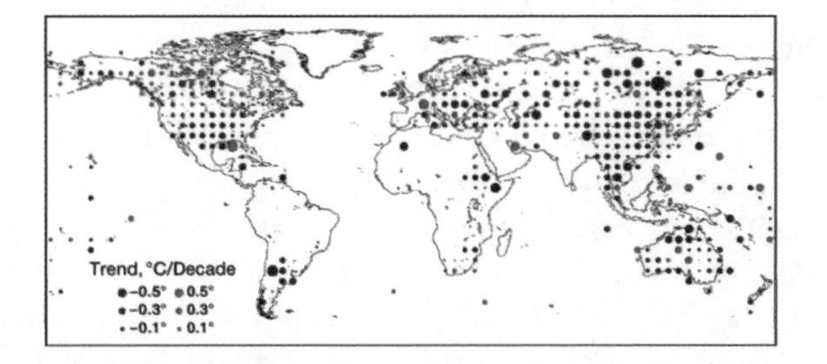

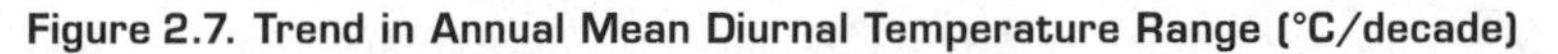

Figure 2.7. Trend in Annual Mean Diurnal Temperature Range (°C/decade)

The magnitude of the trend is given by the area of the circle and the sign of the trend is positive (warming) if the circle is red and negative (cooling) if the circle is blue. Data are from the period 1950–1993 and from non-urban stations only. *Source*: IPCC, 2001.

Other observed changes

It is not possible to provide an exhaustive discussion of all the observed changes in climate since the start of the twentieth century, so we refer the reader to Figure 2.8. This figure captures the essence of most of the major changes observed in the twentieth century. It is important to note that the observed changes are internally consistent with each other, as well as with physical intuition, without needing to appeal to complicated coupled atmosphere-ocean models: increasing greenhouse gases provide a positive radiative forcing that warms the surface of the earth and melts back glaciers, snow, and sea ice. Change is much smaller over the oceans than over land, and hence around Antarctica relative to the Arctic, due to the high heat capacity of the surrounding ocean. A warmer atmosphere holds more moisture so cloud coverage should increase, leading to a reduction in the diurnal temperature range. The hydrological cycle should also intensify, leading to enhanced precipitation at mid-to-high latitudes, with more extreme events and enhanced evaporation at low latitudes.

Figure 2.8 is particularly useful for comparison with Figure 2.15 (next section), which summarizes what a variety of coupled atmosphere–ocean general circulation models (GCMS) project for a future climate warmed through an increase in greenhouse gases. It will be evident that what has already occurred is consistent with what models suggest should have occurred, and also what these same models project will occur more noticeably in the future.

Projections of Climate Change with Applications to Canada

Coupled atmosphere–ocean GCMS have evolved considerably over the years and are continually being further improved, both in terms of resolution and through the inclusion of new, sophisticated, physical parametrisations. These models consist of an atmospheric component, developed through decades of research in numerical weather prediction around the world, coupled to interactive ocean and sea ice models. All GCMS include a land surface scheme, and some now allow the terrestrial vegetation to respond to a changing climate. Climate models are not used to predict weather but rather the slow mean change of average weather and its statistics. They are built on the physical principles that we believe govern the various components of the climate system. Before a climate model is deemed useful for future climate projections, it must be satisfactorily tested against the present-day and transient twentieth-century climate. GCM simulations of past climates (e.g., 6,000 and 21,000 years ago) are also used to evaluate a model's performance against paleo reconstructions. Model deficiencies found through this evaluation process are documented, and attempts are then made to reduce or eliminate them.

Scenarios of future emissions

Any projection of future climate change fundamentally requires assumptions about future emissions of greenhouse gases and aerosols. These in turn are determined by making assumptions about future economic and population growth, technological change, energy use, etc. Clearly, it is

a) Temperature Indicators

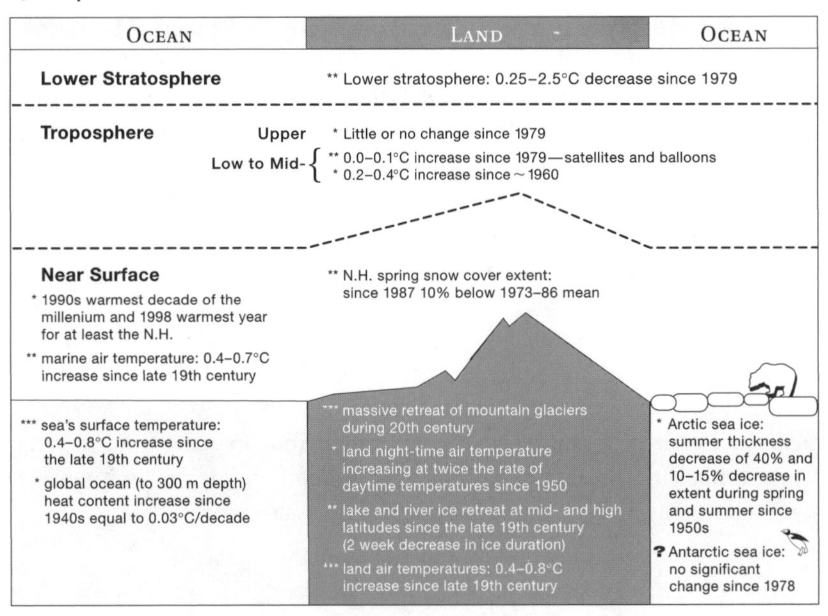

b) Hydrological and Storm-Related Indicators

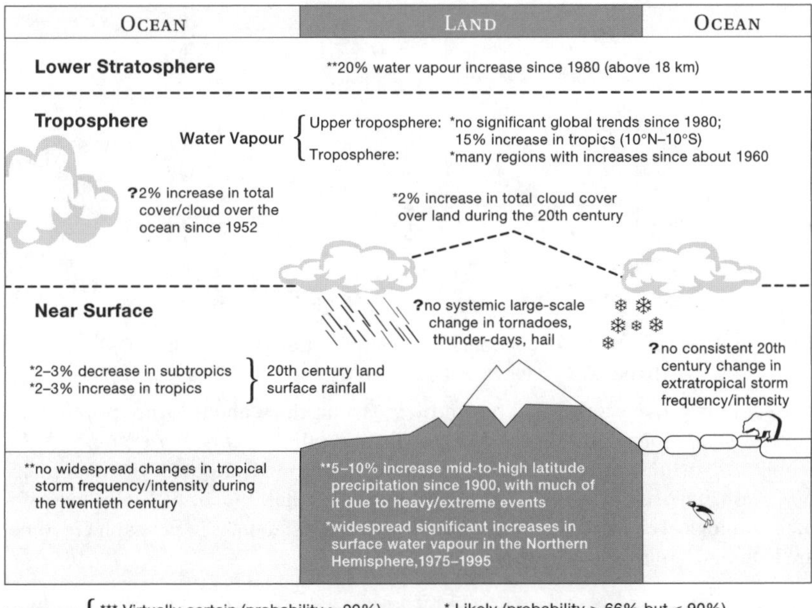

Likelihood { *** Virtually certain (probability > 99%) · ** Very likely (probability > 90% but ≤ 99%) · * Likely (probability > 66% but < 90%) · ? Medium likelihood (probability 33% but ≤ 66%)

Figure 2.8. Schematic Diagram of Observed Variations:

in a) Temperature; b) Hydrological and Storm-Related Indicators. *Source*: IPCC, 2001.

difficult if not impossible to make accurate projections of these socio-economic factors over 100 years. The IPCC therefore developed scenarios of future emissions under a wide range of possible 'story lines' of socio-economic and technological change in the future. In its second scientific assessment, 6 such scenarios were developed (IS92a–f). In the third IPCC assessment, 40 different scenarios were presented (IPCC, 2000).

Several of the possible scenarios lead to projected reductions of greenhouse gases over the twenty-first century (see, for example, the green curve in fig. 2.9). Several other scenarios predict continued growth in emissions, and still others suggest that emissions will continue to increase in the short term but eventually start to decrease. Six sample profiles are shown in Figure 2.9 together with the IS92a or "best guess" profile used in the second scientific assessment. None of the individual scenarios can be termed "correct" since each is equally plausible or implausible. In order to produce a range of possible climate outcomes, it is therefore advisable to integrate coupled models under as many of the scenarios as possible.

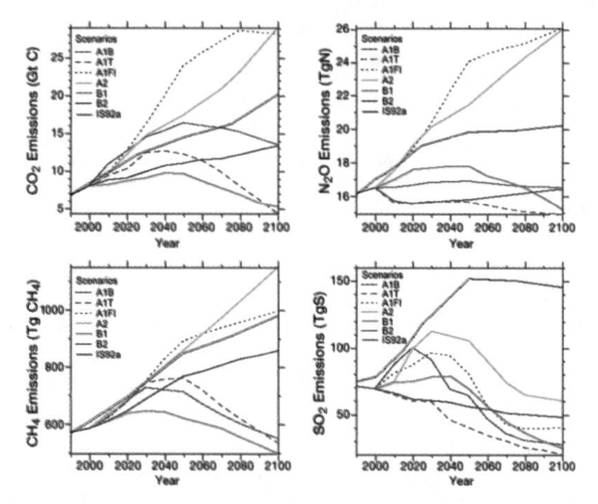

Figure 2.9. Projected Anthropogenic Emissions of Carbon Dioxide (CO_2), Methane (CH_4), Nitrous Oxide (N_2O) and Sulphur Dioxide (SO_2)

The six illustrative scenarios differ in their assumptions about future population growth, technology paths, economic growth, and other factors. For example, A1F1 represents a world characterized by rapid economic growth although global population peaks in the middle of this century and drops after that. A fossil-fuel-intensive path of technological change and a reduction in regional differences in per capita income are assumed. *Source*: IPCC, 2000.

It has, however, not yet been possible for comprehensive GCMs to run century-long integrations using all scenarios because of the lack of available computing time. Nevertheless, several groups around the world have run their GCMs under a select number of these scenarios. The climate sensitivity (defined as the equilibrium warming for a doubling of atmospheric CO_2) and oceanic heat uptake, obtained from the coupled models, can also be used in simpler models to span the full range of scenarios and produce estimates of first-order quantities like global sea-level rise and surface air temperature changes over the next century. These simple models, however, do not allow projections of regional changes in climate.

Projections of future temperature change from simple models

As an initial illustration of the projected global mean surface temperature change over the twenty-first century, we provide Figure 2.10, derived from a simple climate model that uses the climate sensitivity and oceanic heat uptake from more complex models. Using the range of climate sensitivities from coupled GCMs and all emissions scenarios, we arrive at a range of projected 2100 warming, relative to 1990, of 1.4–5.8°C. This range, reported in the IPCC Third Assessment Report, is higher than the 1.0–3.5°C range

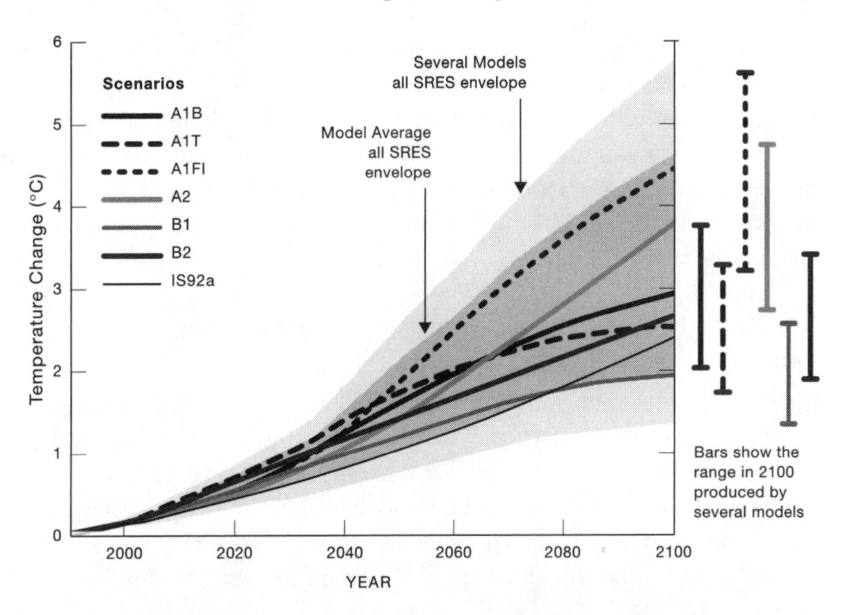

Figure 2.10. Simple Model Projected Global and Annual Mean Temperature Change Relative to 1990 under a Wide Range of Scenarios

The dark shading gives the range using all scenarios and the average model climate sensitivity. The light shading extends this range by calculating the spread from each model independently. Note that in all cases, warming is projected, even though in B1 emissions of CO_2 and CH_4 are assumed to drop substantially below 1990 levels (fig. 2.9). *Source*: IPCC, 2001.

reported in the Second Assessment Report simply because a greater range of scenarios is now being used. That is, in the 1996 report, only 6 scenarios were used, whereas now 40 scenarios are used. Generally, the newer scenarios yield lower sulphur dioxide emissions (lower right panel of Figure 2.9) and hence less cooling from the resulting direct and indirect effects of the aerosols.

The international media picked up on the differences between the range of IPCC 1996 and 2001 projections. Articles appearing in the *National Post*[4] and the United Kingdom's *Daily Telegraph*[5] stated: "Global warming could happen twice as quickly as previously forecast over the next 100 years, the most authoritative report yet produced on the science of climate change said." Similar articles appeared in international, national, and local newspapers, leaving the public with the distinct impression that scientists had now determined that the pace of global warming would double.

The reality, of course, is that in the 2001 IPCC report a wider range of scenarios was used in order to capture a more diverse range of possible socioeconomic assumptions leading to a more diverse range of future emissions. Global warming was not suddenly expected to double (a confusion that arose because the upper bound was raised from 3.5 to 5.8° C); climate models had not become more uncertain (a confusion that arose because the spread between maximum and minimum projected warming had increased from 2.5 to 4.4° C). Very simply, for a more diverse range of emission scenarios, the climate models projected a more diverse range of possible future climates.

Projections of future temperature change from coupled atmosphere-ocean GCMs

As noted above, the enormous computational resources needed to integrate coupled atmosphere-ocean general circulation models under the full 40 scenarios of future emissions meant that only a few illustrative scenarios could be examined for the third IPCC assessment. Figure 2.11 shows the multi-model ensemble mean projected warming averaged over 2071–2100 relative to 1961–1990 for the A2 and B2 scenarios whose emissions are shown in Figure 2.9.

While it is not meaningful to pick a particular place on the earth's surface and say unequivocally that it will warm by a certain amount over the next century, a number of key conclusions can be drawn. First, land areas warm more than ocean areas due to the greater heat capacity of the ocean. Second, the interior of the continents warm more than the coasts as they are further away from the ocean. Third, the west coast at northern mid-

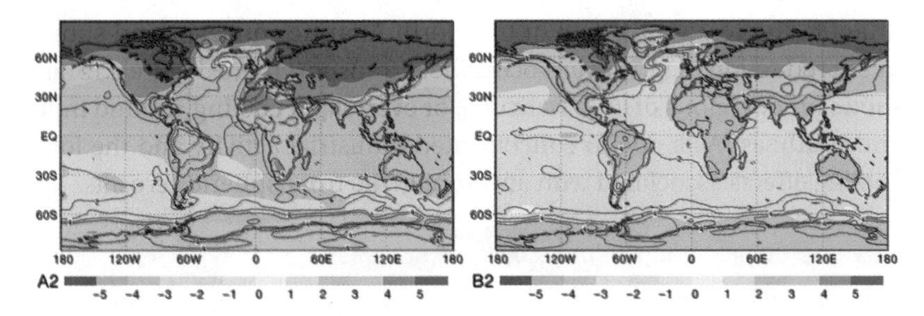

Figure 2.11. The Annual Mean Temperature Change (coloured shading) and Intermodel Range (contour lines) between the 2071–2100 and 1961–1990 Average Climates

Coupled atmosphere-ocean GCMs were driven by either Scenario A2 (left); or scenario B2 (right). All units are in °C. *Source*: IPCC, 2001.

latitudes tends to warm less than the east coast because the prevailing winds are from west to east and thus the west coast is more influenced by the ocean. Fourth, the high latitudes warm more than the lower latitudes due to powerful albedo feedbacks associated with retreating snow and sea ice. As noted above, the retreat of sea ice results in an additional positive feedback since the ocean is no longer insulated from the atmosphere and so can warm it from below. Fifth, the northern hemisphere warms more than the southern hemisphere as there is more land there.

Enhanced warming is also projected in the winter months relative to the summer months, as indicated in Figure 2.12 for the Canadian Centre for Climate Modelling and Analysis model integrated under the IS92a scenario. This particular simulation also reveals local cooling around the North

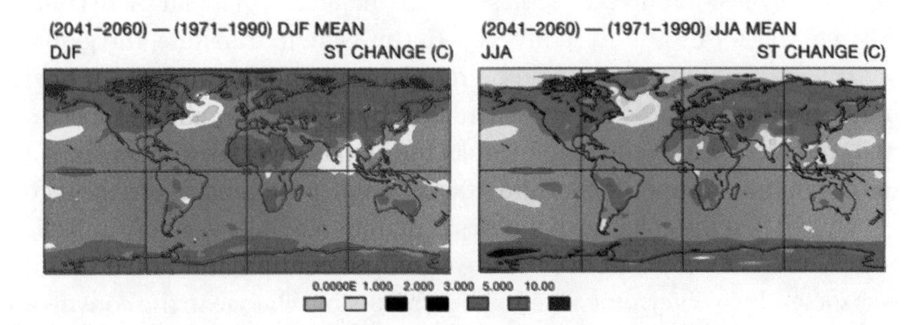

Figure 2.12. Mean Temperature Change between the Average Seasonal Climate in 2041–2060 and 1971–1990 under an IS92 Scenario

Left: Winter: December, January, February; Right: Summer: June, July, August. All units are in °C. *Source*: Dr. G. Flato, Canadian Centre for Climate Modelling and Analysis in Victoria, British Columbia.

Atlantic due to a weakening of the North Atlantic conveyor and subsequent reduction in northward ocean heat transport there. Figure 2.12 also shows other regions of little warming, or even slight cooling, around India and southeast Asia due to concentrated industrial activity and the local cooling effects associated with anthropogenic tropospheric aerosols.

Extreme events and the possibility of surprises

Extreme events

Extreme weather or climate events are important from a policy perspective as they cause the most stress on adaptation strategies for climate change. Adaptation strategies aimed exclusively at dealing with a slow mean change in climate could be ineffective if they do not also account for projected changes in climate and weather statistics associated with the projected mean climate change. In its Third Assessment Report, the IPCC undertook a systematic analysis of observed changes in extreme weather and climate events over the twentieth century and their projected change over the twenty-first century (summarized in Table 2.1, opposite).

Abrupt climate change

Rapid transitions between fundamentally different climate regimes have commonly occurred over the last 400,000 years (fig. 2.1; see Clark et al., 2002 for a review), inspiring scientists to try and grapple with their possible likelihood of future occurrence. Two specific climate change surprises have been given special attention. The first involves trying to determine the probability of a collapse of the West Antarctic ice sheet—an event that would lead to a 6-metre global sea-level rise over a relatively short period of time. The second involves assessing the likelihood of a complete shutdown of the North Atlantic conveyor—if this were to transpire, the global oceanic deep circulation would be reorganized and the amount of heat transported northward in the North Atlantic by the ocean would be substantially reduced; this would tend to affect the climate over land downwind of the ocean (i.e., Europe). In its Third Assessment Report (IPCC, 2001), the IPCC concluded that the collapse of the West Antarctic ice sheet was very unlikely (1–10% chance) to occur over the twenty-first century and noted that it was too early to determine whether an irreversible change in the conveyor is likely or not over this same period.

Most, but not all, coupled model projections of the twenty-first-century climate show a reduction in the strength of the North Atlantic conveyor with increasing concentrations of greenhouse gases (IPCC, 2001). Nevertheless, all coupled model simulations show that Europe continues to warm even in those simulations where the conveyor slows down. In those

Table 2.1 Estimates of Confidence in Observed and Projected Changes in Extreme Weather and Climate Events

Confidence in Observed Changes (latter half of the 20th century)	**Changes in Phenomenon**	**Confidence in Projected Changes (during the 21st century)**
Likely	Higher maximum temperatures and more hot days over nearly all land areas	Very Likely
Very Likely	Higher minimum temperatures, fewer cold days and frost days over nearly all land areas	Very Likely
Very Likely	Reduced diurnal temperature range over most land areas	Very Likely
Likely, over many areas	Increase of heat index (a measure of human discomfort) over land areas	Very Likely, over most areas
Likely, over many northern hemisphere mid- to high latitude land areas	More intense precipitation events	Very Likely, over many areas
Likely, in a few areas	Increased summer continental drying and associated risk of drought	Likely, over most mid-latitude continental interiors (lack of consistent projections in other areas)
Not observed in the few analyses available	Increase in tropical cyclone peak wind intensities	Likely, over some areas
Insufficient data for assessment	Increase in tropical cyclone mean and peak precipitation intensities	Likely, over some areas

Source: IPCC, 2001.

Note: Virtually certain (> 99% chance that a result is true);
Very Likely (90–99% chance);
Likely (60–90% chance);
Medium Likelihood (33–66% chance);
Unlikely (10–33% chance);
Very Unlikely (1–10% chance);
Exceptionally Unlikely (< 1% chance).

simulations where the conveyor reduces in the short term, the ocean acts as a negative feedback to high latitude warming. A reducing conveyor reduces high latitude ocean heat transport and hence sea surface temperatures. This affects atmospheric surface temperatures directly and also affects them indirectly through feedbacks on ice areal extent. Over the longer term, most climate models find a re-establishment of the conveyor to present-day levels: during this re-establishment phase, the ocean conveyor would act as a positive feedback to warming in and around the North Atlantic. What is even less known, and still an outstanding question, is how the stability of the conveyor will change in a future climate warmed through anthropogenic greenhouse gases.

Projected climate change and Canada

Chapter 10 of the IPCC Third Assessment was charged with assessing regional climate information both in terms of the evaluation of regional climates and the projection of regional climate change. This chapter formally showed that the warming found in several coupled atmosphere-ocean GCMs driven by two illustrative scenarios (A2 and B2; see figs. 2.9 and 2.11), was 40% above the global average in the winter months at high northern latitudes (see fig. 2.13). The eastern North American (ENA), central North American (CNA), and western North American (WNA) regions, which include most of southern Canada, showed greater than average warming both in summer and winter in both the A2 and B2 scenarios. For the eastern Arctic/Greenland (GRL) region, which includes some of northern Canada, greater than average warming in the summer and much greater than average warming in the winter months are projected. The 1.4–5.8°C globally averaged warming projected by 2100 should be considered to be amplified over most of Canada (see figs. 2.11 and 2.12).

Similarly, there is intermodel agreement that the GRL and Alaskan (ALA) regions, which include much of northern Canada and all of the Canadian Arctic, will receive at least a 5%–20% increase in precipitation in summer and winter by the year 2071–2100 (fig. 2.14). Under the A2 scenario (figs. 2.9 and 2.11) greater than 20% increases are projected for these regions. Increases of 5%–20% in precipitation by 2071–2100 in the winter are projected for both the WNA and ENA regions, although in the summer as well as in the CNA region, intermodel differences are of inconsistent sign. The impact of these projected changes in regional precipitation will be discussed further in chapter 4.

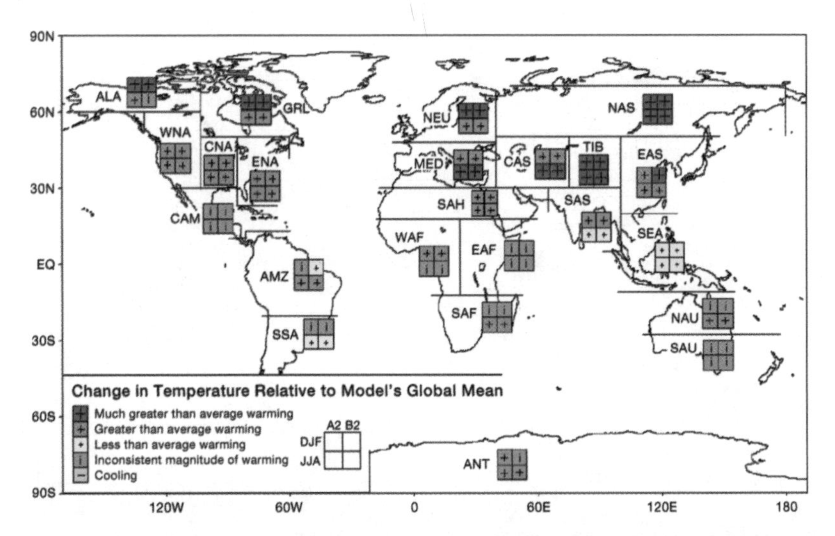

Figure 2.13. Analysis of Inter-Model Consistency in Regional Relative Warming (relative to each model's global average)

Regions are classified as showing either agreement on warming in excess of 40% above the global average (much greater than average warming), agreement on warming greater than the global average (greater than average warming), agreement on warming less than the global average (less than average warming), or disagreement amongst models on the magnitude of regional relative warming (inconsistent magnitude of warming). There is also a category for agreement on cooling (which never occurs). A consistent result from at least seven of the nine models is deemed necessary for agreement. The global annual average warming of the models used spans 1.2–4.5°C for A2 and 0.9–3.4°C for B2, and therefore a regional 40% amplification represents warming ranges of 1.7–6.3°C for A2 and 1.3–4.7°C for B2. *Source*: IPCC, 2001.

Summary

There are simply too many projected changes from too many coupled atmosphere-ocean GMCs to provide a comprehensive description of all aspects of future projected change. Figure 2.15, from the IPCC Third Assessment, summarizes the projected changes for the end of the twenty-first century with an assigned level of confidence (virtually certain: > 99% probability; very likely: 90–99% probability; likely: 66–90% probability; medium likelihood: 33–66% probability). The comparison of Figure 2.15 with Figure 2.8 suggests that what has already occurred in the climate record is consistent with what models suggest should have occurred, and also with what these same models project will occur more noticeably in the future.

Certain phenomena listed in Figure 2.8 are not listed in Figure 2.15 as they are either not resolved, or there is a disagreement between models as to what might occur. Tornadoes, thunder days, hail, and lake and river ice

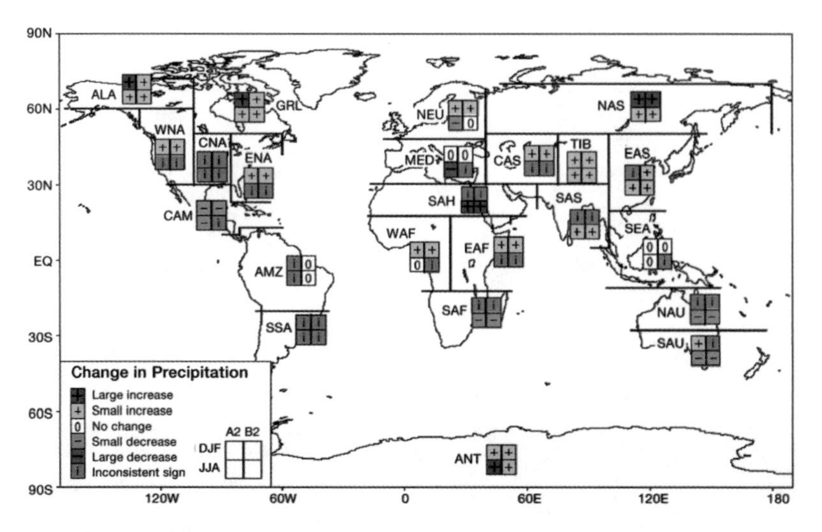

Figure 2.14. Analysis of Inter-Model Consistency in Regional Precipitation

Regions are classified as showing either agreement on increase with an average change of greater than 20% (large increase), agreement on increase with an average change between 5 and 20% (increase), agreement on a change between −5 and +5% or agreement with an average change between −5 and 5% (no change), agreement on decrease with an average change between −5 and −20% (decrease), agreement on decrease with an average change of less than −20% (large decrease), or disagreement (inconsistent sign). Increases or decreases are for the 2071–2100 mean relative to the 1961–1990 mean. A consistent result from at least seven of the nine models is deemed necessary for agreement. *Source*: IPCC, 2001.

melt, for example, are not resolved in coarse resolution climate models so no assessment can yet be made as to their future changes. Climate models all have cloud parametrizations that differ from model to model and the resulting changes in amount and type of clouds varies between models.

Climate Change Detection and Attribution

The climate system varies on a variety of time scales both through *natural*, internal processes as well as in response to variations in *external* forcing (e.g., solar changes, volcanic emissions, greenhouse gases). As such, the detection of climate change involves looking, in a statistically significant sense, for the emergence of a signal above the background of *natural* climate variability. Attribution involves specifically assigning a cause for the detected signal to human activities, variations in other external forcing, or a combination of both.

a) Temperature Indicators

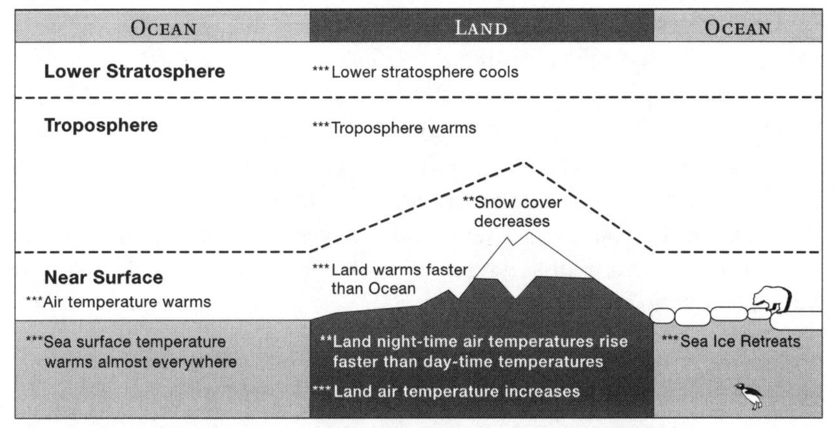

b) Hydrological and Storm-Related Indicators

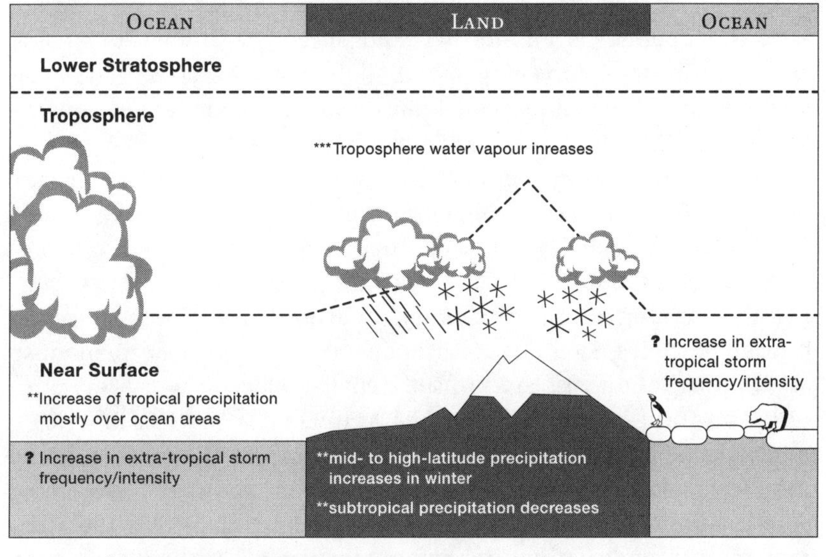

Likelihood
- *** Virtually certain (many models analyzed and all show it)
- ** Very likely (a number of models analyzed show it, or change is physically plausible and could readily be shown for other models)
- * Likely (some models analyzed show it, or change is physically plausible and could be shown for other models)
- ? Medium likelihood (a few models show it, or results mixed)

Figure 2.15. Schematic Diagram of Variations in a) Temperature; b) Hydrological and Storm-Related Indicators

Source: IPCC, 2001.

In 1996, the IPCC second scientific assessment stated, "The balance of evidence suggests a discernible human influence on global climate." This cautious statement represented the consensus view of those scientists working in WGI. It was based on the results of only a few detection and attribution studies available at the time. Since 1996, there have been many more published works firming up the science behind this statement. As a result, the third WGI IPCC assessment, released in January 2001, made a much stronger statement: "There is now new and stronger evidence that most of the warming observed over the last 50 years is attributable to human activities" (IPCC, 2001).

In fact, there does not appear to be any detection and attribution study that has been able to explain the warming in the second half of the century through any known natural cause. Global mean surface air temperatures have increased by 0.6 ± 0.2°C since 1860, although this warming has not occurred in a constant fashion. Most of the warming has occurred during two distinct periods, from 1910–1945 and since 1976, with a very gradual cooling during the intervening period. Global warming critics have been quick to point out that most models that have simulated the climate of the twentieth century have failed to capture this feature.

Researchers at the Hadley Centre in the United Kingdom have recently reported upon the most comprehensive simulations to date of the climate of the twentieth century (Stott et al., 2000). They found that natural forcing agents (solar forcing and volcanic emissions), while necessary to simulate the warming early in the century, could not account for the warming in recent decades. Similarly, anthropogenic forcing alone (greenhouse gases and sulphate aerosols) was insufficient to explain the 1910–1945 warming but was necessary to simulate the warming since 1976 (see fig. 2.16).

By regressing the large-scale signals from their simulations on decadal mean observations, Stott et al., (2000) demonstrate that natural forcing alone is not a plausible explanation for the observed changes in the twentieth century, and that natural and anthropogenic forcing both make significant contributions to the observed change. They show that when combined, these signals explain approximately 80% of the observed interdecadal variance of global mean temperature.

The experiments performed by Stott et al., took into account estimated historical variations in the main anthropogenic and natural external forcing agents that are believed to have affected the climate of the past century. These include heat-trapping greenhouse gases, changes in ozone abundance (also a greenhouse gas), and the formation of sulphate aerosols from

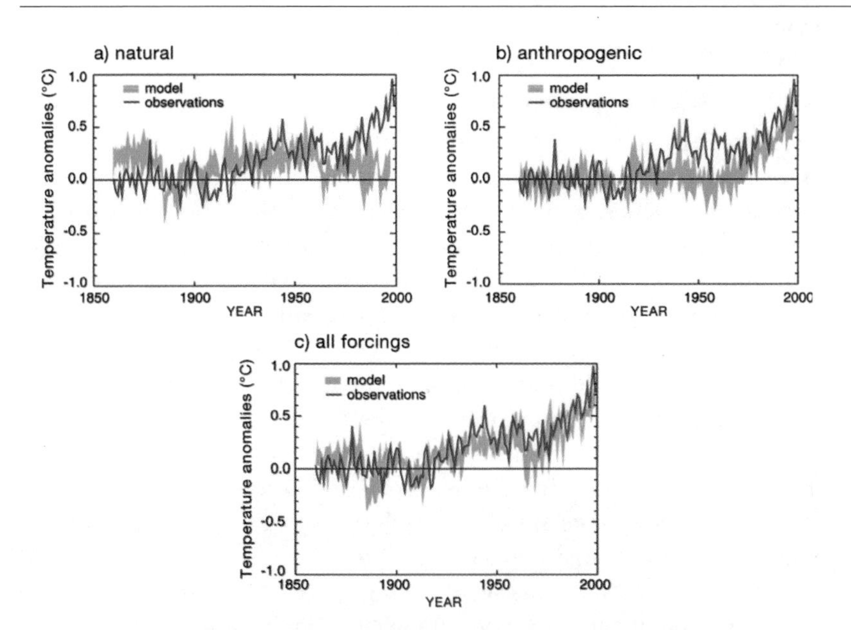

Figure 2.16. Annual and Global Mean Temperature Anomalies (relative to the 1880–1920 average) from a suite of climate simulation experiments with both natural and anthropogenic forcing agents

The gray shading indicates the range spanned by four independent simulations; the red line indicates observations: a) natural forcing (solar + volcanic emissions) experiments produce a gradual warming to about 1960 followed by a return to late nineteenth-century temperatures, consistent with the gradual change in solar forcing throughout the twentieth century and a resumption of volcanic activity during the past few decades; b) the anthropogenic (greenhouse gas + aerosols) simulations reproduce the warming of the last three decades, but underestimate the early century warming and do not capture the slight cooling that occurred between the two periods of rapid warming. The combined forcing simulations, however, are able to reproduce much of the observed decadal scale variation in global mean temperature and are also able to capture with some fidelity the large-scale spatial structure of the observed changes. *Source*: IPCC, 2001.

the industrial emission of sulfur dioxide. Their approach was far from a diagnostic curve-fitting exercise. Rather, a model built on physical principles was used to simulate the climate's response to independent estimates of historical climate forcing. The striking level of agreement obtained between observed and simulated decadal scale temperature variations strongly supports the contention that radiative forcing from anthropogenic activities, moderated by variations in solar and volcanic forcing, has been the main driver of climate during the past century.

Science and Policy

In this chapter I have outlined the 200-year history and science of global warming. In particular, I have pointed out that the science is deeply rooted in the peer-reviewed literature. Unfortunately, the media tend to sensationalize the science, often leaving the general public confused. An example of this comes from the *Victoria Times Colonist,* which on 14 January 2001 published a leading article titled, "Study deflates global warming." On 23 January 2001, only nine days later, the lead story on the front page was headlined, "Global warming severity grows." The average person reading these pieces would think that scientific understanding was swinging like a pendulum from one extreme to the other. They would be entirely confused, and might even dismiss the whole issue, since they would not have the benefit of reading the peer-reviewed literature from which the stories arose. Scientists are of course debating climate change in peer-reviewed, international journals. Yet this debate is not about *whether* human-caused climate change is taking place—but rather about how quickly, to what magnitude, and with what regional implications.

In focusing my discussion, I appealed to the findings of the latest IPCC report, which states, "There is now new and stronger evidence that most of the warming observed over the last 50 years is attributable to human activities." This warming trend, and its seasonal and large scale-geographical distribution, are consistent with what coupled models have suggested should have occurred, and also what these same models project will occur more noticeably in the future.

With respect to Canada, Figure 2.5d shows that there has been a strong warming trend in annual mean temperatures, especially during winter (fig. 2.6a). Projections of future climate change all consistently show that warming will continue in the region, under all scenarios of future emissions which have been used to drive the coupled models. Most of Canada is projected to have greater than average warming in the summer and winter, for both the A2 and B2 scenarios. Northern Canada is, however, projected to have greater than average warming in the summer but much greater than average warming in the winter months, with precipitation increases occurring in both seasons. The 1.4–5.8°C globally averaged warming projected by 2100 should therefore be considered to be amplified over Canada.

Uncertainty in climate change projections

There are two types of uncertainties involved in climate change projections (NRC, 2003): those that are essentially random (aleatory uncertainty), and those that arise from an incomplete knowledge or understanding of a particular process (epistemic uncertainty). By its very definition, aleatory uncertainty is impossible to reduce (i.e., the odds of getting "heads" when flipping a coin once is never less than 50%). In the context of climate change, the estimation of aleatory uncertainty is often accomplished by using one model to create an ensemble of climate model integrations. The range spanned by the different integrations, which differ only in their initial condition, then represents an estimate of uncertainty associated with random processes and natural climate variability. The mean of the ensemble represents a best estimate (e.g., see fig. 2.16).

An estimation of epistemic uncertainty can be obtained by using different models with different parametrizations of unresolved processes and integrating each of these with the same radiative forcing. Thus the intermodel variation gives an estimate of model uncertainty and the intramodel variation gives an estimate of uncertainty associated with natural variability.

The epistemic uncertainty in climate change projections can be further broken down into two components: one involving uncertainty in climate feedbacks, and one involving uncertainty in the emissions scenario used to drive the climate model. In terms of overall uncertainty, each contributes about 50%, the latter being dependent on poorly constrained assumptions of future population growth, social behaviour, economic growth, energy use, and technology change. Compounding the problem of uncertainty is the potential existence of 'unknown unknowns' whose importance only becomes apparent once they are discovered.

Extensive research has been conducted over the last several years in an attempt to quantify uncertainty in climate change projections. Stott and Kettleborough (2002) found that in the absence of policies to mitigate climate change, climate change projections over the next 40 years are insensitive to the particular emission scenario used (see also Zwiers, 2002). Knutti et al., (2002) further found that there is a 40% chance that actual warming at 2100 will exceed the upper bound of the range (1.4–5.8°C) estimated in the IPCC Third Assessment Report. They found only a 5% chance that it will be less than the lower bound.

While science is likely to reduce the epistemic uncertainty of the known unknowns over the next decade, it is also likely to discover new unknowns. In terms of projections of climate change over the next 40 years, it is unlikely that science will change the global estimate and range of warming. Nevertheless, where science is likely to make substantial reduction in uncertainty is with regards to the regional downscaling of climate projections.

Improved regional projections will allow the development of local adaptation strategies while the necessary international negotiations to move towards significantly reduced global emissions takes place.

Future directions of climate modelling

In the IPCC Third Assessment Report, none of the international groups contributing projections of future climate had incorporated interactive terrestrial and oceanic carbon cycle models into their coupled model. Several international groups have subsequently made significant advances in this regard, and the first projections including interactive carbon cycle and dynamic terrestrial vegetation are beginning to appear. A major thrust of international coupled modelling efforts over the next few years will be the development of a terrestrial and oceanic carbon cycle modelling capability for use in climate change projections on which policy will be based.

In the IPCC fourth assessment, likely to occur in 2007, the leading climate models will include interactive terrestrial and ocean carbon cycles in which anthropogenic greenhouse gas and aerosol emissions, rather than concentration scenarios, will be specified. In addition, it is likely that these same models will allow both vegetation type and function to vary with the changing climate, thereby taking account of important biological feedbacks within the climate system. The state-of-the-art climate models will also incorporate interactive atmospheric sulphur and ozone (tropospheric and stratospheric) cycles, which will allow a more complete treatment of natural and anthropodenic radiative forcing of the climate system.

In the IPCC fifth assessment, probably in ten years time, it is likely that, rather than specifying future emissions of atmospheric aerosols and greenhouse gases, the state-of-the-art models will calculate these emissions internally through the interaction of coupled climate/socio-economic models. That is, emissions will be calculated internally under specified policy, technological, population growth, and other socio-economic options.

Scientific challenges

Some degree of climate change is inevitable as the Earth system moves towards a new global radiative equilibrium under increased levels of greenhouse gases. While the Kyoto Protocol represents small step towards addressing the issue of global warming, it is only a start. In fact, if we wish to stabilize atmospheric CO_2 levels at 3–4 times preindustrial values, global anthropogenic emissions must be reduced to less than half of what they were in 1990 (IPCC, 2001). Meeting the required global reductions in fossil-fuel emissions presents numerous scientific challenges, yet at the same time offers an enormous economic opportunity. Our ability to reach and maintain a yet-to-be-defined acceptable level of climate change will require the spawning of new energy technologies (see chap. 6). The market for these technologies is global, the field is wide open, and every single individual on this planet is a potential customer.

In addition to cutting CO_2 emissions at their source, through both changes in lifestyle (chap. 12) and the development of new technologies (chap. 6), enhancing natural carbon sinks may prove to be a viable and cost-effective, if only short term, approach to mitigation (see chap. 5). At the same time, regional adaptation strategies need to be developed for the change that is already in the pipeline (chap. 8). The science behind the projections of regional climate change is still in its infancy. There are large uncertainties, both in downscaling global climate change projections onto a sub continental scale, and in the basic physics of processes (such as clouds and precipitation) that operate on these regional scales.

Concluding Remarks

Over the course of the twentieth century, the globe has warmed by $0.6 \pm 0.2°$C as a consequence of natural and anthropogenic changes in radiative forcing. If greenhouse gas emissions were to be curtailed immediately, such that the atmospheric concentration of CO_2 remained fixed at 368 ppm, the Earth would still warm a further $\sim 0.5°$C over the next several centuries (Weaver et al., 2001) as it tries to equilibrate to the increased radiative forcing. This warming follows from the oceans' slow response to perturbations in radiative forcing. This said, if emissions were to be stopped, natural carbon sinks, especially in the ocean, could potentially draw down and store a significant portion of past emissions.

In a recent study, Ewen et al., (2004) found that the ocean had the capacity to take up an additional 65–75% of the atmospheric CO_2 increase when anthropogenic forcing was stopped. This decreased by about 5% for each 50 year period that anthropogenic emissions were maintained at a stabilized and elevated atmospheric greenhouse CO_2 level. The results of their work clearly have enormous and encouraging policy implications with respect to future fossil-fuel emissions. If we are able to reduce emissions in the near future, there is hope that the ocean can draw down a substantial portion of the atmospheric CO_2 solely through the solubility pump (the dissolution of CO_2 in water); 65–70% of all past emissions can be drawn down into the ocean. The longer it takes to reduce emissions, the less CO_2 the ocean solubility pump will be able to draw down. Nevertheless, uncertainties in natural carbon cycle feedbacks are large and are only just beginning to be examined by the scientific community.

Much as the Kyoto Protocol requires countries to consider the implications of their greenhouse gas emissions beyond their immediate national borders, reducing the uncertainties in the science of global warming requires scientists to transcend traditional disciplinary boundaries to meet the challenges raised in this chapter. Meeting these challenges will create new scientific opportunities, and from these opportunities we will determine what is an acceptable level of future change.

Notes

1 For example, in Global Warming: The Clouds Thicken, by Peter Foster, *National Post*, 19 August 2000; Global Warming Fears Cool Off. Why Impose Questionable Constraints on Economic Growth, by Peter Holle, *Winnipeg Free Press*, 27 January 2001.

2 For example: US National Academy of Sciences, National Research Council, *CO_2 and Climate: A Scientific Assessment* (1979) and *Changing Climate* (1983).

3 The albedo of a surface is defined as the percentage of incoming solar radiation hitting the surface that is reflected back to space.

4 *Hot, Hot Is the Range*, by Margaret Munro, 23 January 2001.

5 *Pace of Global Warming "Could Double,"* by Charles Clover, 25 January 2001.

References

Arrhenius, S. (1896). On the Influence of Carbonic Acid in the Air upon the Temperature of the Ground. *London, Edinburgh and Dublin Philosophical Magazine and Journal of Science* 5th Ser.: 237–76.

Christianson, G.E. (1999). *Greenhouse: The 200-Year Story of Global Warming*. New York: Walker.

Christy, J.R., D.E. Parker, S.J. Brown, I. Macadam, M. Stendel, and W.B. Norris. (2001). Differential Trends in Tropical Sea Surface and Atmospheric Temperatures Since 1979. *Geophysical Research Letters* 28: 183–86.

Clark, P.U., N.G. Pisias, T.F. Stokcer, and A.J. Weaver. (2002). The Role of the Thermohaline Circulation in Abrupt Climate Change. *Nature* 415: 863–69.

Ewen, T.L., A.J. Weaver, and M. Eby. (2004). Response of the Inorganic Ocean Carbon Cycle to Future Warming in a Coupled Climate Model. *Atmosphere–Ocean*, in press.

Fourier, J.B.J. (1824). Remarques générales sur la température du globe terrestre et des espaces planétaires. *Annales de chimie et de physique* 27: 136–67.

IPCC. (1996). *Climate Change 1995: The Science of Climate Change*. Contribution of Working Group I to the Second Assessment Report of the Intergovernmental Panel on Climate Change. Cambridge, UK: Cambridge University Press.

IPCC. (1998). Principles Governing IPCC Work. <http://www.ipcc.ch/about/princ.pdf>.

IPCC. (2000). *Emissions Scenarios.* A Special Report of Working Group III of the Intergovernmental Panel on Climate Change. Cambridge, UK: Cambridge University Press.

IPCC. (2001). *Climate Change 2001: The Scientific Basis.* Contribution of Working Group I to the Third Scientific Assessment Report of the Intergovernmental Panel on Climate Change. Cambridge, UK: Cambridge University Press.

IPCC-WGI. (n.d.) Working Group I: The Science of Climate Change. Hadley Centre for Climatic Prediction and Research. <www.meto.gov.uk/research/hadley centre/ipcc/wg1/home.html>.

Knutti, R., T.F. Stocker, F. Joos, and G.-K. Plattner. (2002). Constraints on Radiative Forcing and Future Climate Change from Observations and Climate Model Ensembles. *Nature* 416: 719–23.

Manabe, S., and R.T. Weatherald. (1967). Thermal Equilibrium of the Atmosphere with a Given Distribution of Relative Humidity, *Journal of the Atmospheric Sciences* 24: 241–59.

National Research Council. (1979). *CO_2 and Climate: A Scientific Assessment.* Washington, DC: National Academy Press.

National Research Council. (1983). *Changing Climate.* Washington, DC: National Academy Press.

National Research Council. (2000). *Reconciling Observations of Global Temperature Change.* Washington, DC: National Academy Press.

National Research Council. (2003). *Understanding Climate Change Feedbacks.* Washington, DC: National Academy Press.

Prabhakara, C., R. Iacovazzi, Jr., J.-M. Yoo, and G. Dalu. (2000). Global Warming: Evidence from Satellite Observations. *Geophysical Research Letters* 27: 3517–20.

Revelle R., and H. Suess. (1957). Carbon Dioxide Exchange Between the Atmosphere and the Ocean and the Question of an Increase of Atmospheric CO_2 during the Past Decades. *Tellus* 9: 18–27.

Stott, P.A., and J.A. Kettleborough. (2002). Origins and Estimates of Uncertainty in Predictions of Twenty-First Century Temperature Rise. *Nature* 416: 723–26.

Stott, P.A., S.F.B. Tett, G.S. Jones, M.R. Allen, J.F.B. Mitchell, and G.J. Jenkins. (2000). External Control of Twentieth Century Temperature by Natural and Anthropogenic Forcings. *Science* 290: 2133–37.

Weaver, A.J., M. Eby, E.C. Wiebe, C.M. Bitz, P.B. Duffy, T.L. Ewen, A.F. Fanning, et al., (2001). The UVic Earth System Climate Model: Model Description, Climatology and Application to Past, Present and Future Climates. *Atmosphere-Ocean* 39: 361–428.

Wigley, T.M.L. (1998). The Kyoto Protocol: CO_2, CH_4 and Climate Implications, *Geophysical Research Letters* 25: 2285–88.

Zwiers, F.W. (2002). The 20-year Forecast. *Nature* 416: 690–91.

The Human Challenges of Climate Change

Steve Lonergan

Introduction

THE RAPIDLY CHANGING CLIMATE detailed by Andrew Weaver in the previous chapter has been the result of a broader human transformation of the natural environment. The human causes of this transformation are varied and have as much to do with institutional and cultural norms and practices as with economic and social ones. In this chapter on the human challenges of climate change, I introduce two types of activities that are primarily responsible for this transformation: (1) changing land use; and (2) increased fossil fuel consumption. Subsequently, I review the potential human consequences of climate change and outline the response options that have been developed and/or discussed to either mitigate the causes or adapt to the consequences. The remainder of the chapter discusses the social processes and driving forces that represent the true human challenges of dealing with climate change, with specific emphasis on equity, and offers some options for the future.

It is clear that the causes of climate change, the consequences, and the options for response vary widely over space. To date, developed countries have been largely responsible for the increase in atmospheric concentrations of carbon dioxide (CO_2). Since 1950, five countries (US, Russia, Germany, Japan, and the UK) account for well over 50% of the total CO_2 emissions globally (ORNL, 2001; Table 3.1). The US alone contributed almost 30% of the total emissions from fossil fuel combustion. Canada's contribution to total global emissions of CO_2 was only 2.1% in 1996, but it also exhibits one of the largest per capita emission levels in the Organization for Economic and Co-operative Development (OECD), behind only the US and

Australia (see fig. 3.1). Immediately, this brings into question issues of equity in dealing with the consequences of climate change and developing appropriate response options.

Table 3.1. Cumulative Emissions of CO_2 since 1950, Top 12 Countries

Country	Cumulative CO_2 Emissions, 1950–1995 (1000 tonnes)	% of Total Emissions (since 1950)
United States	180,245,575	27.31
Russia	66,694,682	10.11
China	54,030,802	8.19
Germany	41,784,828	6.33
Japan	29,736,951	4.51
United Kingdom	26,666,955	4.04
Ukraine	20,934,158	3.17
France	16,443,057	2.49
India	14,507,388	2.20
Canada	14,467,674	2.19
Poland	14,009,231	2.12
Italy	11,924,026	1.81

Source: ORNL, 2001.

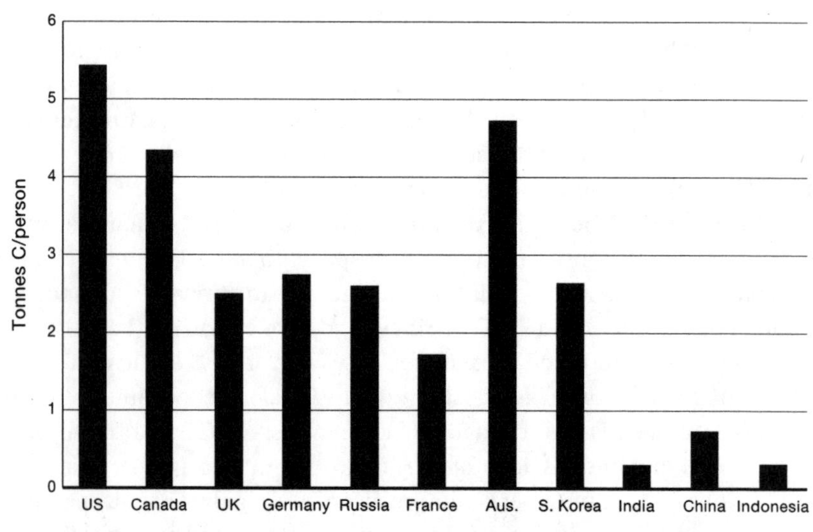

Figure 3.1. Per Capita Emissions of Carbon, 1998 (selected countries)

Source: ORNL, 2001.

While emissions of CO_2 and other greenhouse gases vary widely from country to country, so do the consequences of climate change. Developed countries may experience changes in water supply, biodiversity, and crop production, but the countries least able to accommodate a changing climate are the poorest countries of the South. The most recent report from the Intergovernmental Panel on Climate Change (IPCC) is very clear: "The impacts of climate change will fall disproportionately upon developing countries and the poor persons within all countries, and thereby exacerbate inequities in health status and access to adequate food, clean water, and other resources" (2001, p. 7). Poor economic conditions make these people and their communities vulnerable to any type of environmental disruption, and their ability to respond—part of their vulnerability—is extremely limited. The human challenges of climate change are closely tied to the inequitable distribution of causes, consequences, and response options available. Like the issue of sustainable development, equity is a key principle, and much of this book addresses this issue.

What Are the Major Human Causes of Climate Change?

Land use and land-cover change

The role of human activities in changing the land cover is central to the study of environmental change. Almost 20% of the world's forests have been converted to cropland and pastureland over the past 300 years (Richards, 1990). The alteration of the land cover and changes in the way land is used affect the biogeochemical cycles of the Earth, the level of atmospheric greenhouse gases, and other land surface characteristics (see Table 3.2).

Table 3.2 Changes in Land Cover, Globally, 1700–1980

Land Cover Type	Area in 1700 (millions of hectares)	Area in 1980 (millions of hectares)
Forest and Woodlands	6,215	5,053
Grassland and Pasture	6,860	6,788
Croplands	265	1,501

Source: Richards, 1990; National Research Council, 1999.

Changes in land cover are primarily the result of human use (Allen and Barnes, 1985; Turner et al., 1990). Human use of the land involves both the ways in which land is manipulated, and the intent underlying that manipulation (LUCC, 1998). Land use change is influenced by driving forces, such as population growth and development, which in turn are affected by existing environmental conditions, economic systems, culture, and other socioeconomic and behavioural attributes. This makes modeling the links between land use change and climate change complex and difficult.

The challenge for society is to understand better how changes in land use affect all aspects of the biophysical system and to minimize the environmental impacts associated with land manipulation. In some countries (such as Brazil and Indonesia), land-use change—primarily the rapid clearing of forest land—contributes enormous amounts of CO_2 to the atmosphere (see chap. 5). However, land cover can act as both a sink and an emitter of carbon dioxide and other greenhouse gases. Appropriate land management can offset at least some of the increases in greenhouse gases from fossil-fuel consumption. Accordingly, land use, land-use change, and forestry (LULUCF) activities play an important role in international policy being developed to address climate change.

Energy and material use

In the past, CO_2 emissions have been closely tied to fossil-fuel consumption which, in turn, has been linked to economic output. Since 1950, total global energy consumption has increased from 76,500 petajoules (10^9 J) to almost 400,000 petajoules (or 2,610 million tonnes of oil equivalent [Mt oe] to over 12,000 Mt oe), and per capita consumption has increased by 20% as well (Darmstadter, 1971). Increased global economic output poses enormous challenges because of the apparent link between economic output, energy consumption (from fossil fuels), and carbon dioxide emissions. Fossil fuels account for roughly 90% of the world's commercial energy consumption. Global energy use is expected to increase to almost 15,000 Mt oe by the year 2010, with most of the increase due to economic growth in Asia. Energy use in Canada has increased by 25% since 1985, from 7,523 PJ to 9,404 PJ, and is expected to reach 15,000 PJ (roughly 422 Mt oe) by 2020 (EIA, 2001). Canada is also the highest consumer of energy per dollar of gross domestic product of any country in the OECD. Admittedly, cross-national comparisons are somewhat fallacious; geography, resource availability, and history may play a more important role than efficiency of production. Sector specific studies illustrate the energy inefficiency of Canadian industry (Lonergan, 1999). However, it is clear that the growth rates in global energy consumption (approximately 2% per year) could be tempered with aggressive policies to promote energy efficiency and fuel substitution; estimates of future emissions beyond 2010 are quite sensitive to changing economic, technological, and social systems as well as to the institutional context that is presently being developed. Growth in energy consumption and CO_2 emissions is expected to be particularly rapid in China and South Asia, due to industrial expansion, population growth and urbanization, and increased per capita incomes. Exacerbating the problem is the fact that coal, which produces the highest CO_2 emissions of any fossil fuel, is the major source of electricity for China (70%) and South Asia (60%), and electricity demand in these regions is rising at close to 7% per year.

For some countries, such as Canada, the link between CO_2 emissions and economic growth was slightly "decoupled" during the late 1970s and early 1980s. Because of technological change, less carbon is emitted per dollar of economic output than was the case 40 years ago (from 1.2 tonnes of $CO_2/\$10^6$ of output in 1958 to approximately $0.8/\$10^6$ in 1998). Figure 3.2 shows the growth in CO_2 emissions along with the growth of the national economy. Before 1978, CO_2 emissions grew steadily with (and even more rapidly than) economic output. Over the succeeding 8 years, the relationship between emissions and GDP appears to have weakened (or been "decoupled"); but the past 15 years have seen a "recoupling" of CO_2 emissions and GDP, highlighting the difficulty Canada faces in reducing greenhouse gas emissions and maintaining economic growth. As well, the absolute level of CO_2 emissions continues to rise, despite the country's

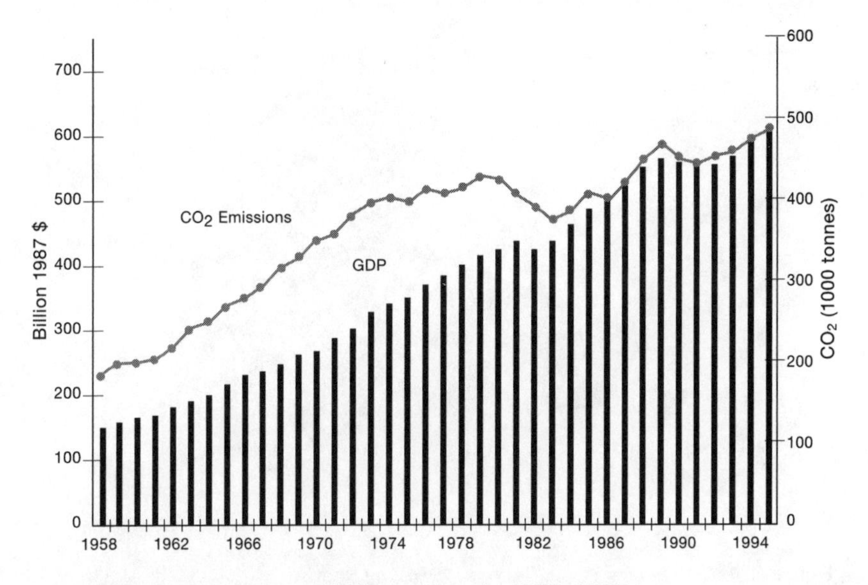

Figure 3.2. CO_2 Emissions and GDP for Canada, 1958–1995

Source: ORNL, 2001; Statistics Canada.

commitment under the Kyoto Protocol to decrease emissions by 6% (from 1990 levels) by the year 2012. One of the major human challenges of climate change will be to "decarbonize" the global economy, starting with developed countries. In other words, we need to find a way to maintain a reasonable level of global economic output while reducing our level of fossil-fuel energy consumption.

Canada signed the Kyoto Protocol in 1997 and ratified it in 2002, thereby agreeing to reduce its CO_2 emissions by 6% of 1990 levels. However, CO_2 emissions were 13.5% higher in 1998 than in 1990 (Environment Canada, 2000), and they continue to rise.

At present, two sectors in Canada account for more than half of the greenhouse gases: transportation and power generation (see fig. 3.3). Industry accounts for an additional 15% of emissions. Any attempt to reduce Canada's emissions will have to focus on these three sectors. However, meeting Canada's commitments to reduce greenhouse gas emissions through a reduction in domestic emissions is not politically palatable. Most Canadian reductions will occur by paying for efficiency improvements in other countries, through mechanisms incorporated into the Kyoto Protocol.

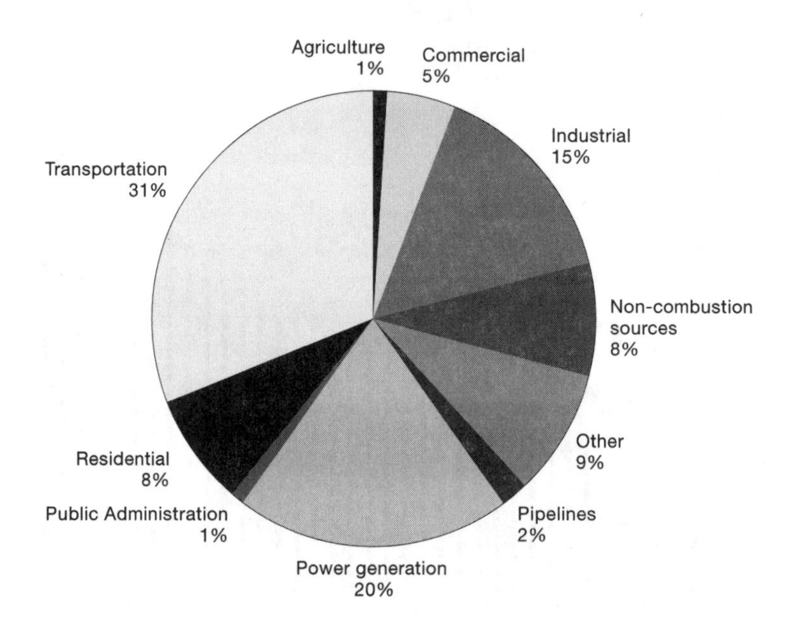

Figure 3.3. CO_2 Emissions in Canada, 1995, by Sector

Source: Environment Canada, 1998.

What Are the Human Consequences of Climate Change?

Introduction

A recent report by the Royal Society of Canada concluded that the greatest impacts of climate change on Canada would likely be on boreal forests (see chap. 5), water supplies, coastal ecosystems, and human health (Royal Society, 2001). Some of the problems include increased fire and pest outbreaks in our boreal forests, higher heat wave mortality in major cities, melting of large areas of permafrost, and disruptions in water supply in the Great Lakes and the Prairies. In the discussion below, I highlight impacts in four areas: human security, health, water resources, and food systems.

Human security

Resource scarcity and environmental degradation create inequities (or the perception of inequities) in resource distribution that often contribute to insecurity, and conflicts and tensions over natural resources such as energy and freshwater have been well documented (Westing, 1986; Homer-Dixon, 1991; Lonergan and Brooks, 1994). More difficult to determine, and possibly more devastating, are the long-term and somewhat diffuse impacts on national and human security that may result from climate change. Admittedly, the term "human security" is itself somewhat diffuse; but for our purposes, human security is achieved when and where individuals and communities

- have the options necessary to end, mitigate, or adapt to threats to their human, environmental, and social rights;
- have the capacity and freedom to exercise these options; and
- actively participate in attaining these options.

Human security is, accordingly, a term that encompasses many of the human dimensions of global change: health, resource availability, vulnerability to hazards, and environmental degradation (as well as crime, drugs, and assorted other problems). Climate change affects human security since it may have significant implications for resource availability, agricultural productivity, and economic output, and may lead to coastal flooding and the creation of "environmental refugees." Sea-level rise, now projected to be between 0.2 and 0.6 metres under a scenario of doubling CO_2 levels, will have significant impacts on low-lying regions and countries such as Egypt and Bangladesh, where a large percentage of both the population and the productive capacity is located less than one metre above sea level. More disruptive to political stability, however, are the increasing magnitude and frequency of extreme events—events that are difficult and costly to prepare for and that may cause major social disruption. Most concerned will be those regions that are most vulnerable to climate disruptions, particularly areas subject to floods and droughts.

Coastal flooding, a constant problem in much of Southeast Asia, is expected to increase both in terms of flood frequency and magnitude. This could cause population displacement and the related problems of environmental refugees. Periodic droughts in arid and semi-arid regions, already a cause of population displacement and conflict, could become more frequent and more long lasting. As noted above, the recent IPCC report acknowledged that the greatest impact of climate change and associated extreme events would be on those groups in society that are most vulnerable to external stresses—the disenfranchised and impoverished who exist in all countries. The United Nations Development Programme (UNDP, 2000) recently estimated that over one billion people live in absolute poverty in the developing world, 64% of them in Asia. Since many impoverished populations also live in ecologically fragile areas, climate change could be devastating to such groups. The impacts of climate change—biophysical, socioeconomic, and political—as well as the present discussions on response strategies, must be considered in the context of the relationship between poverty and environmental change.

The inability to adapt to rapid or extreme changes in climate is a major challenge for human systems. Abrupt changes may also force ecological and human systems beyond important thresholds. It seems clear that biophysical systems may be able to withstand a certain level of stress (high water levels, for example). At some point, the system exceeds its capacity to withstand this stress (and, using the previous example, river banks may be flooded). Our understanding of similar thresholds for social systems is much less developed. Higher temperatures could trigger adverse health impacts or have a negative economic impact on certain sectors which, in turn, could affect the entire economy. Another key challenge is to better understand these social thresholds and how climate change might affect them (NRC, 2002).

In terms of human security, the negative consequences of climate change on a particular sector or issue (such as the availability of fresh water) is not the only concern. The possible synergistic or cumulative effects of these impacts also require study. A decrease in global food production, a decrease in water availability, the potential for longer and more severe droughts, and land degradation through erosion, flooding, and pollution (e.g., salt water intrusion) will act together to reduce the ability of certain communities to grow enough food. However, the main cause of malnutrition—presently affecting almost one billion people in the world—is not simply insufficient food production. The main cause is poverty, which, as a main component of human security, is likely to be exacerbated by climate warming (World Resources Institute, 1998).

Health

Climate change is expected to adversely affect human health both directly (through increased heat wave mortality and loss of life in extreme events such as floods, storms, or sea-level rise) and indirectly (through changes in existing health risks, such as inadequate water, lack of food, or an increase in vector-borne and water-borne diseases). Malnutrition is one of the biggest global health problems, and it contributed to approximately 12% of all deaths in 1990 (World Resources Institute, 1998). This is a case where the various potential negative consequences of climate warming may combine to increase what is already a major health concern.

Much of the discussion about the human health impacts associated with climate change has focused on the potential increase in infectious diseases worldwide. Temperature and precipitation affect the abundance and distribution of disease vectors and disease-causing microbes, which, in turn, are likely to affect disease rates. The most obvious examples are diseases that are caused by mosquitoes, such as malaria, dengue fever, yellow fever, and certain types of encephalitis (McMichael et al., 1996). Increases in temperature would extend mosquito ranges into temperate regions or to higher altitudes. Higher temperatures also affect the transmission rate of diseases; a 2° c increase in temperature cuts the incubation period for the dengue virus almost in half, with a resulting threefold increase in transmission rate of the disease (World Resources Institute, 1998). Precipitation affects the availability of breeding sites for mosquitoes, so changes in disease incidence are likely to vary widely over space. Temperature increases also influence the spread of algae blooms, the habitat for the microbe that causes cholera. As Eyles and Sharma (2001) note, major outbreaks of cholera as a result of climate warming. Table 3.3 summarizes some of the factors that are likely to be influenced by climate change and the types of diseases that may, in turn, be affected. One of the key challenges of climate change is thus not only the risk to human health, but also the increasing burden on health services. Questions about the ability of health service institutions to adapt and respond to new and increased threats raise serious concerns about the impacts of climate change (see chap. 8).

Table 3.3 Climate Change and Disease Risk: Types of Diseases that Might Be Affected

Factor	Examples of Specific Factors	Examples of Diseases
Ecological changes (including those due to economic development and land use)	Agriculture; dams, changes in water ecosystems; deforestation/ reforestation; flood/drought; famine; temperature increases; precipitation increases	Schistosomiasis (dams); Rift Valley fever (dams, irrigation); Argentine hemorrhagic fever (agriculture); Hantaan (Korean hemorrhagic fever) (agriculture); hantavirus pulmonary syndrome, Southwestern US, 1993 (weather anomalies); malaria (climate), dengue (climate)
Human demographics behaviour	Population growth and migration (including environmental refugees and movement from rural areas to cities); war or civil conflict; urban decay	Introduction of HIV; spread of dengue; spread of HIV and other sexually transmitted diseases

Source: Adapted from Eyles and Sharma, 2001.

As with many of the consequences of climate change, the distribution and spread of infectious diseases have a disproportionate effect on the poor. Poverty and the increased risk and incidence of infectious disease are mutually reinforcing. Again, the inequitable distribution of the consequences of climate change poses major challenges in planning future responses.

Water resources

The impact of climate change on water resources will vary widely from region to region. At present, almost 2 billion people, or one-third of the world's population, live in countries that are considered water stressed. The general notion of "water stress" and "water scarce" as developed (with different methods of measurement) by Falkenmark et al., (1989), Gleick (1996), and Ohlsson (1999) are widely used, but they are not particularly useful. Not only are there various forms of water stress and scarcity, but the level of stress is a function of many social, economic, and institutional factors, such as efficiency of use, level of national income, and level of agricultural development. Nevertheless, it is clear that as population grows, the number of countries experiencing water shortages will increase. At present, over two dozen countries exhibit conditions of water stress or water scarcity; by 2025, this number will rise to almost 50, and by 2050, it is expected to reach 54, with a total population of over 4 billion persons (Falkenmark et al., 1989; Gleick, 1996). Climate warming on these levels of stress will likely add to the problem. Climate change models are consistent in projecting less rainfall for much of Africa, the Mediterranean, and Australia, regions that are presently experiencing water stress. Since irrigation accounts for over 70% of water use, higher demand for water from the growth in food production, loss of soil moisture, and the increase in evaporative demand will worsen water stress in many regions.

More important might be the increased magnitude and frequency of floods and droughts. The IPCC (1996) notes that the flood-related consequences of climate change may be equal to, or greater than, the impacts of droughts. Some regions may experience both of these phenomena as the variability in climate increases. Since many of our water storage facilities and water management plans were based on past climatic and hydrological parameters, the potential impact of climate change must be taken into consideration in future water resource system design.

Food systems

The impact of climate change on food systems and food security has probably received the most interest over the past two decades. Changes in crop yields resulting from climate change are expected to vary widely by region, depending on crop type, soil moisture, and a variety of other factors. For small increases in temperature, the consequences are expected to be positive or, at the worst, neutral, as farmers are able to adapt to changing conditions. For larger increases (above 2° C), there will likely be negative consequences, including increased heat stress, decreased yields, increased prices, and increased severity of droughts (IPCC, 2001). As is the case for other systems, the ability to adapt to changes will affect the severity of impacts. Agroindustry in North America and Europe will be able to adjust planting dates, type of crops, fertilization rates, and the like, to ensure a minimal, or even positive, impact of climate change. Farmers in developing countries, however, do not have a similar range of options. Crops in tropical countries may be sensitive to even small temperature changes, and in these regions the ability to adapt is limited. More important might be the negative impact of lower commodity prices (from increased production in the North) on agriculture-based economies. With higher increases in temperature, food prices would likely increase as the food supply would be constrained (IPCC, 2001). Again, there would be an effect on the incomes of poor urban consumers and on populations vulnerable to drought.

Possible Consequences for Canada

The impacts of climate change on Canada have been studied quite extensively, and a detailed summary can be found in the Canada Country Study that was published by Environment Canada (Koshida and Avis, 1998 and the IPCC, 2001).[1] Chapter 8 covers in detail many of the impacts on Canada, and the purpose here is simply to introduce some of the more severe impacts that are expected. Although Canada has enormous water resources, there is a significant regional variation in availability and accessibility. Climate change is expected to have significant impacts on some aspects of water availability, including decreases in lake levels, groundwater levels, and soil moisture. While precipitation may increase for most regions of the country, this will be more than offset by increases in evapo-transpiration, resulting in greater demand for irrigation water. Lower lake levels could cause a reduction in hydroelectric power production and in the amount of available fish habitat. The demand for water for irrigation and domestic use will likely increase. However, warmer temperatures may also result in a longer shipping season, particularly in the North, with attendant economic benefits—albeit with possible environmental and sovereignty risks (Lonergan et al., 1993).

The impacts of climate change on agricultural production in Canada have been well studied (see Koshida and Avis, 1998), but, like the global impacts noted above, projections are linked to the level of expected change in temperature and precipitation. Longer frost-free seasons will occur, along with potentially higher precipitation. However, the increase in evapo-transpiration noted above will lead to moisture deficits across the country (Koshida and Avis, 1998). The additional CO_2 in the atmosphere could potentially have a positive effect on crop yields, further complicating the issue. Most of the impact studies have focused on agriculture in Ontario and the Prairies, and they conclude that the economic impacts will be relatively small (Brklacich et al., 1998). Last, the impact of global changes in production and commodity prices will also affect Canada, as our export markets change. Again, the extent of the impact is uncertain, and could range from expanded markets for Canadian wheat to lower prices, which would adversely affect production.

One of the key sectors in Canada is the forest industry, supporting one in every 15 jobs in the country. Large areas of forest decline are expected to result from climate change. While the northern range of individual species may increase, it is believed that the climate will change more rapidly than the forests can adapt. Losses due to increased fires, pest outbreaks, and drought stress will have negative impacts on the industry.

The impacts of climate change on human health in Canada will depend on how Canadians are able to adapt to higher temperatures and potentially higher humidity. Heat-related deaths would likely increase; estimates run as high as 1000 additional deaths for both Montreal and Toronto (Koshida and Avis, 1998). There would likely also be a higher incidence of respiratory disorders and an increase in the transmission of vector-borne diseases. However, as in the case of agricultural productivity, the possibility of adaptation to these changing conditions will reduce the negative impact of climate change on human health.

Human Responses to Climate Change

International environmental agreements

Much of the work on responses to climate change has focused on the development of multilateral environmental agreements (MEAS). These include the Montreal Protocol (ozone depletion), the Biodiversity Convention, and the UN Framework Convention on Climate Change (UNFCCC) and its Kyoto Protocol. In 1992, representatives from most of the countries in the world met in Rio de Janeiro to adopt a framework for international action on the issue of the increasing concentrations of greenhouse gases in the atmosphere. At the meeting, it was felt that efforts must be made to reduce both the level of emissions and the overall concentration of greenhouse gases in the atmosphere. The resulting framework—the UNFCCC—provided an initial step toward this emission reduction goal. The UNFCCC has now been ratified by over 180 countries. It entered into force in March 1994, and has since been supplemented with the Kyoto Protocol (agreed to at a meeting in Kyoto, Japan, in December 1997), which identified specific emission reduction targets for 38 developed countries (the so-called Annex B countries, since these commitments are listed in Annex B of the Protocol). Some of these countries, along with the goals agreed to at Kyoto, are listed in Table 3.4. The Protocol represents a binding agreement to reduce the global emission of greenhouse gases. To take the force of international law, the Protocol must be ratified by 55 countries representing over 55% of the total greenhouse gas emissions globally. The goal is to reduce greenhouse gas emissions from developed countries by 5% by the year 2012. To date, countries representing 44% of global emissions have ratified the Protocol; how-

Table 3.4. Annex B Countries and Their Kyoto Targets.

Country	Kyoto Target (% change from 1990 emissions)	Country	Kyoto Target (% change from 1990 emissions)
Australia	+8	Poland	−6
Canada	−6	Romania	−8
EU	−8	Russian Federation	0
Hungary	−6	Switzerland	−8
Iceland	+10	Ukraine	0
Japan	−6	US	−7
New Zealand	0	Hungary	−6
Norway	+1		

Source: Kyoto Protocol to the United Nations Framework Convention on Climate Change.

ever, it will only come into effect as international law if either the US or Russia ratify, thereby surpassing 55% of global emissions.

The reduction of greenhouse gas emissions can be accomplished in two ways: either by reducing the rate at which CO_2 (and other greenhouse gases) is added to the atmosphere (this generally occurs through fossil-fuel combustion or biomass burning); or by increasing the rate at which CO_2 is removed from the atmosphere (by somehow storing additional carbon in the biosphere). The former could be accomplished through improved efficiency of energy production and end use, or by reducing the carbon content of fuels through a combination of decarbonization, fuel switching, and increased use of non-carbon energy systems (e.g., renewables and nuclear energy). The latter approach involves reducing net emissions through sequestering carbon by enhancing natural carbon sinks (e.g., increased forestation) or by capturing and storing CO_2 that has been emitted from fossil-based energy systems (e.g., in deep geologic formations or in the ocean). For economic and technological reasons, it is unlikely that Canada, the US, or a number of other countries can meet their CO_2 emission reduction commitments through improved energy efficiency and a decarbonization of their respective economies. However, it is conceivable, and even desirable, that these commitments could be met by the large-scale removal of carbon. The sequestration option has been broadly identified as land management activities, or Land Use, Land Use Change, and Forestry (LULUCF). LULUCF could include, for example, decreasing the amount of carbon lost or emitted through deforestation, or increasing the rate of uptake of carbon from the atmosphere by afforestation or reforestation. It could also include materials substitution: substituting renewable biomass fuels for fossil fuels (e.g., forest energy plantations). The net direct impact on a region would then be the sum of all of these mechanisms.

Economic costs and ratification

One of the biggest concerns of the United States, Canada, Australia, and a few other developed countries is the economic cost associated with adopting any greenhouse gas mitigation strategy. For the US, these costs have been estimated as high as 3% of Gross Domestic Product (GDP) (see chap. 7). Offsetting these costs are the direct and indirect benefits that would accrue from emission reduction. First, reducing emissions will limit the damages expected from climate change. The IPCC estimates that the economic costs of stabilizing emissions would range from −0.5 percent of GDP (for OECD countries) to +2.0 percent of GDP. Greater energy conservation alone could result in substantial reductions in greenhouse gas emissions at zero net economic cost.

More important, might be the indirect benefits from emission reduction. The IPCC notes that "double dividends" could result as efforts to reduce greenhouse gas emissions would also reduce other pollutants. The indirect economic benefits could offset 30% or more of mitigation costs (IPCC, 2001). Further, if the emission reduction strategy includes an energy or carbon tax, there income will be generated to offset other costs. Despite numerous studies showing that the costs of mitigation would be minimal or even negative, the US and other countries have adamantly refused to ratify the Kyoto Protocol.

The Key Challenges: Social Processes and Driving Forces

A number of the human challenges of climate change were noted above in the discussion about consequences and responses. Developing suitable adaptation mechanisms, promoting the decarbonization of the economy, and determining how impacts and responses in one region may affect other regions are all major concerns. In this next section, I focus on some of the specific "driving forces" that directly impact greenhouse gas emissions and, therefore, the extent of climate change that we can expect. This is followed by a discussion of the key challenge I believe we face—the issue of equity.

Driving forces

One key force driving greenhouse gas emissions has been the rapid globalization of economies and cultures—the major socio-economic change of the past century. The development of transnational corporations that wield considerable economic and political power, the emergence of global markets for goods, the increased mobility of capital and labour, and the conversion of socialist economies to a market-based structure have transformed the globe, with significant implications for the natural environment. This has resulted in both greater risks and improved opportunities to confront the issue of climate change. The rapid growth of fossil-fuel-based economies in India and China could dramatically increase the level of greenhouse gas emissions (fig. 3.4 shows the case for India; the graph for China has the same shape). Both countries—along with most other developing nations—have resisted international attempts to constrain their economies for the sake of the global environment. However, the development of multilateral environmental agreements, such as the UNFCCC noted above, has provided opportunities to address these risks while at the same time decreasing greenhouse gas emissions in developed countries.

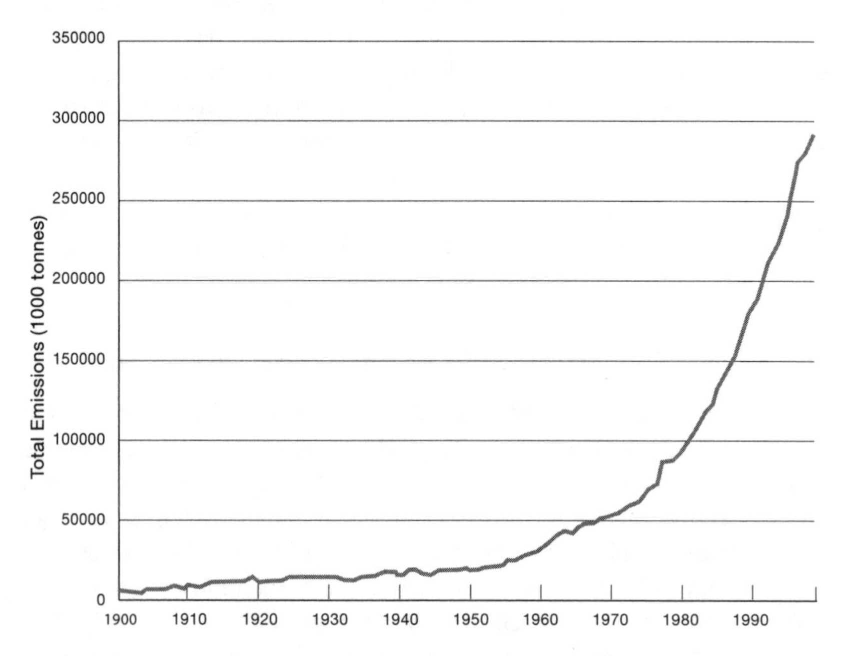

Figure 3.4. India CO_2 Emissions, 1900–1998 (from fossil fuels)

Source: ORNL, 2001.

To ensure sustainability, this global economic transformation will have to incorporate environmental concerns. Production and consumption systems must become less dependent on fossil fuels, and the development of alternative energy systems and environmentally benign technology must be a high priority for all countries. Our major challenge with respect to climate change is to ensure that these considerations are included in all international agreements—including trade agreements—and that nations discontinue subsidizing systems that pose threats to the global environment. Canada must not only meet its commitments to the Kyoto Protocol, but also reassess its reliance on fossil-fuel production and consumption.

A second driving force will be technological change, which is discussed in chapter 6. Decarbonization and dematerialization—that is, producing goods with less carbon emissions and material throughput—will be an important part of this transformation process. Technological change has already improved the energy efficiency of production in most countries and is a major reason for the decline in the amount of energy (and, therefore, carbon) used per dollar of economic output in countries like Japan, Germany, and Sweden (ORNL, 2001). One of the key components of the Kyoto Protocol is the transfer of "clean" technologies from the developed to the developing world, through what is known as the Clean Development Mechanism (CDM). The CDM will allow countries like Canada to meet their carbon emission reduction commitments by transferring clean technology to developing countries, thereby reducing the latter's production of carbon.

How quickly can technological change be effected? And how important will fuel substitution and improvements in efficiency be in reducing our greenhouse gas emissions? There is little doubt that technological change could have a significant impact on carbon emissions. However, many factors influence the energy and carbon intensity of an economy, and there is much debate about the role technological change and/or demand management does, and should, play. However, there is general agreement that these factors must be integrated into any long-term strategy to reduce greenhouse gas emissions.

A third driving force is public perception about and attitude toward the existence of climate change and what to do about it. Although the scientific evidence outlined by Andrew Weaver in the previous chapter is clear and the scientific consensus on climate change overwhelming, the media has portrayed the issue as divisive and fraught with many uncertainties. There has also been much discussion of the potential economic costs of greenhouse gas mitigation. As a result, public attitudes in North America have been ambivalent about how politicians should proceed. There is strong support for improved environmental quality across most countries and most socio-economic groups, but despite this public concern, there is little support for major agreements that would impose economic costs on society, particularly with respect to global issues such as climate change.

Canada has now ratified the Kyoto Protocol, and we have stated publicly that we will honour our commitments to reduce CO_2 emissions (from 1990 levels) by 6%. The US, however, has made no such claim, and the US Congress has made it clear that China and India must be part of any climate change agreement before the US would be willing to even consider ratification. President Bush has also made it clear that the US rejects the Kyoto Protocol and will not be a party to it. Key opponents to the climate change agreement come from energy-producing states (and energy-producing provinces in Canada) and from those convinced that there will be economic hardships from any agreement that will affect people and nations unequally. While there seems to be more support for the climate change agreement in Canada than in the US, both countries continue to increase fossil-fuel production and consumption.

Fourth, institutions are important driving forces affecting all aspects of the natural and human environment. Inadequate institutional structures can cause environmental problems, such as resource depletion or land degradation. However, institutions are also essential to solve environmental problems, and the UNFCC and the Montreal Protocol are two notable examples involving global climate issues. As Young (1999) notes, "the operation of institutions accounts for a sizable proportion of the variance in human behaviour affecting biogeophysical systems." (For a complete discussion of the past and future of climate change institutions see chaps. 10 and 11.)

Institutions can operate at different spatial levels, from the local (e.g., water management) to the international (e.g., the Kyoto Protocol). In this chapter, I have focused on international institutions such as the UNFCCC and its Kyoto Protocol because of the global nature of climate change. Agreements must be made at the international level to coordinate emission reduction activities. However, local institutions can have an important impact as well. Examples might include using market-based options to reduce vehicular travel, promoting public transit, or subsidizing the use of alternative energy vehicles. There is little question that one major human challenge of climate change is designing institutions that work—that is, institutions that are efficient and effective, and that incorporate enforcement.

Two institutions are central to discussions on climate change: the first is the UNFCCC, discussed above; the second is the Intergovernmental Panel on Climate Change (IPCC) which was developed by the World Meteorological Organization (WMO) and United Nations Environmental Program (UNEP) in 1988. In the previous chapter, Weaver has outlined the structure of the IPCC. The role of the IPCC is to assess the scientific, technical, and socio-economic information relevant for understanding the risk of human-induced climate change. It does not carry out new research nor does it monitor climate related data. It bases its assessment mainly on published and peer-reviewed scientific technical literature.

The Key Challenge: Equity

The concept of equity—defined by the International Court of Justice as a "general principle directly applicable as law"—is a key component of many international treaties. To date, a number of treaties have incorporated the concept of intergenerational equity, including the Vienna Convention and associated Montreal Protocol on Protection of the Ozone Layer. While there are no precise guidelines for defining equity, it has been a central issue in climate change discussions since their inception. The issue is so important that the IPCC organized a workshop on this topic in 1994 (IPCC, 1995).

Statements from the Non-Aligned Movement, the European Parliament, the Africa Group and China over the past three years have all identified global fairness and the assignment of equitable emissions entitlements as key to the acceptance of any climate change agreement (Agarwal et al., 1999). Much of the discussion has been around the principle of "contraction and convergence." That is, the principle of equity demands that all nations agree to the same level of per capita emissions entitlements and work toward achieving this by a specified date. Between the present and this future date, there will be a contraction of both total global emissions of greenhouse gases and per capita emissions in developed countries, while developing countries (in most cases) will be allowed to increase their per capita emissions until there is global convergence. This is no trivial matter, since per capita emissions for most OECD countries vary from 10–20 tons of CO_2/person (in 1995; the numbers are even higher for some OPEC countries) to virtually zero for some countries in sub-Saharan Africa. Per capita emissions for selected developed countries are shown in Figure 3.1, and the variations even within this group are notable. While agreement on contraction and convergence is the crux of the argument about global fairness, the issue of equity appears in other forms in the UNFCCC and its Kyoto

Protocol. For example, one of the key components of the Kyoto Protocol is the Clean Development Mechanism, or CDM, as noted above. The CDM is a mechanism designed to facilitate the transfer of environmentally benign technology from the North to the South. In exchange, northern countries would get credit (toward their own emission reduction commitments) for reducing the emissions in the recipient countries. The CDM has two explicit objectives: first, to reduce the global emissions of greenhouse gases; and second, to promote sustainable development in the South. However, a cornerstone of sustainable development is—or should be—equity. International equity, intergenerational equity, and intra-generational equity are all aspects of equity that must be considered (Lonergan, 1993). Until equity becomes a cornerstone of a future climate change agreement, it is unlikely that we will achieve much-needed reductions in global greenhouse gas emissions or, for that matter, meet the goal of sustainable development that is explicit in the Kyoto agreement.

In the context of the climate change discussions, five aspects of equity must be considered (see Agarwal et al., 1999; Claussen and McNeilly, 1998; and Lonergan, 1993):

- responsibility for the problem;
- the distribution of impacts associated with the problem;
- the ability to pay to mitigate or adapt to climate change;
- the opportunities or options available to make changes; and
- the distribution of benefits from an agreement.

Unquestionably, one of the key challenges we face with respect to climate change is how to integrate these five components of equity.

Responsibility

The first issue is the responsibility for the present state of affairs with respect to greenhouse gas emissions. As noted above, five countries account for more than 50% of CO_2 emissions since 1950. Present emission levels—either in absolute terms or per capita—are similar, except that China now accounts for 12% of total emissions (ORNL, 2001). There is also concern about future responsibility for greenhouse gas emissions, as estimated by the average rate of growth of emissions by country (and taking into consideration the Kyoto targets). The conclusion from this cursory historical analysis is inescapable: developed countries have been primarily responsible for the emission of greenhouse gases into the atmosphere. The only change in this scenario in the future will be the growth of emissions in China and India, and the potential stabilization of emissions—and possibly their reduction—in developed countries.

The assumption here is that those who bear the responsibility for the environmental harm should also pay to remedy the problem (now commonly referred to as the "polluter pays principle" or PPP). While it seems clear that those who bear the responsibility for past CO_2 emissions should pay for both the removal of CO_2 and any damages, the complex nature of climate change (with multiple contributors, unknown damages and enormous emission reduction costs) considerably muddles this discussion.

Distribution of impacts

The uneven distribution of the effects of climate change constitutes a second aspect of the issue of equity. The impacts or consequences of climate change on human systems were outlined above. The most vulnerable to climate change—or any environmental stress for that matter—are the poor in developing countries and the developed world. In terms of the potential human toll, however, people in the South are much more at risk from such impacts as sea-level rise in low-lying island states, longer and more severe droughts in the Sudan, and flooding episodes in Bangladesh. While it is clear that the responsibility for climate change lies primarily with the North, the adverse consequences of climate change will fall disproportionately on the South.

Inequities in income and ability to pay

A third aspect of equity relates to the enormous gap in resources (economic, institutional, and other) between developed countries and developing countries. Per capita GDP figures highlight this difference. If the responsibility for CO_2 emissions falls primarily on the developed nations of the world, it is also clear that the ability or capacity to pay for any reduction in emissions is also highest for these same countries (with the exception of newly independent states, or NIX). There is enormous inequality in the standard of living among countries, and the gap between developed and developing countries is increasing (UNDP, 1998). However, regardless of who bears the responsibility for climate change, or who might be culpable in the future, climate change is global, and the consequences will be global as well as local. The costs of mitigation or adaptation must be borne by those who have the ability to pay.

Opportunities for emission reduction

A fourth aspect of inequity in climate change relates to the opportunities available to reduce greenhouse gas emissions. Inefficient economic sectors (in terms of CO_2 emissions per dollar of output) would be the first targeted for improvement. There are large variations in energy intensity across countries, from China, with 65.9 GJ/$ GDP in 1996, to Japan, which used only 6.7 GJ/$ in the same year (see Table 3.5). The implications of these variations could be enormous on economic and social levels, and within and between countries. Any long-term strategy to address CO_2 emissions reduction will involve a move away from fossil fuels. For energy intensive industries (and regions), this implies greater costs (for fuel substitution, for example) or even dislocation of industry. The present Kyoto Protocol does not require emission reductions from developing countries, so it is likely that, where possible, some energy intensive industries will move to developing countries to avoid the costs of operating in the North. Regions and countries that produce fossil fuels may also be impacted; resource-dependent places like Alberta in Canada or West Virginia in the US may experience negative economic impacts. This will also be true internationally. These differences will result in inequity in the cost of emission reduction, as developed countries face large economic costs, while inefficient industrial and transportation sectors in developing countries offer much cheaper emission reduction alternatives. This issue is recognized in the Kyoto Protocol, and mechanisms for emissions trading and joint implementation are built into the agreement. However, differing opportunities for emission reduction

Table 3.5 Energy Intensity of Selected Countries, 1980–1996 (in gigajoules of energy per dollar of GDP [1990 US$])

Country	1980	1990	% Change 1980–1990	1996	% Change 1990–1996
China	167.0	101.1	−39.5	65.9	−34.8
Germany	13.7	8.6	−37.2	7.2	−16.3
Egypt	33.2	31.8	−4.2	28.6	−10.1
India	60.4	49.6	−17.9	45.9	−7.5
Kenya	72.0	61.2	−15.0	57.6	−5.9
US	17.7	14.5	−18.1	14.2	−2.1
South Africa	30.3	36.8	+21.5	37.1	+0.8
Mexico	17.2	17.8	+3.5	18.1	+1.7
Brazil	12.1	13.0	+7.4	13.3	+2.3
France	8.3	8.0	+3.6	8.3	+3.8
Japan	7.7	6.4	−16.9	6.7	+4.7
Canada	27.3	22.7	−16.8	27.9	+22.9

Source: Agarwal et al., 1999.

must also be viewed in the context of the differing levels of economic well-being as well as the level and type of resources available. As noted above, 70% of China's commercial energy consumption is from coal, its most abundant resource.

Distribution of benefits

Fifth, there will likely be an inequitable distribution of the benefits of mitigation and adaptation. While it would be extremely pretentious to suggest that developing countries should pay for the benefits they reap from mitigating climate change (that was caused by the developed world), some feel that countries that benefit the most from action on climate change should take on a greater share of the burden.

How should these five aspects of equity relative to climate change be incorporated into an eventual agreement? Claussen and McNeilly (1998) offer five principles:

- All nations should be able to maintain or improve standards of living.
- The outcome of the UNFCCC negotiations should not undermine progress toward the goal of sustainable development.
- The countries most responsible for greenhouse gas emissions should be leaders in the effort to reduce emissions.
- All nations should work to the best of their abilities.
- The world should take advantage of emission reduction opportunities when they exist.

Unfortunately, these principles are not that helpful. They merely offer general guidelines and say nothing about how to ensure that equity becomes a cornerstone of any climate change agreement. Agarwal et al., (1999) are much more aggressive in advocating "contraction and convergence" as part of any agreement. While acknowledging that ecological and economic effectiveness is an important objective, they promote the need for equity and global solidarity as equally essential principles. Ensuring that these principles are integrated into an eventual agreement seems to be the main challenge that we face in developing an international agreement.

Many of the concerns about equity fall under the concept of "intergenerational fairness" or "intergenerational equity." One of the themes that appears throughout this book is the principle that regardless of what actions we take today, changes to the world's climate that may occur during our lifetime—and the concomitant impacts—are the result of our past activities. The present concern with the degradation of our atmosphere relates to how we care about *future* generations. A commitment to sustainability and sustainable development implies a commitment to equity with future generations. This issue is further developed in the chapter by Coward toward the end of this book.

The Need for Integrated Assessment

A final challenge we face is the need to develop improved scientific assessment in the area of the human dimensions of climate change. It is impossible to devise effective environmental policy without reliable scientific information and data collection. To date, such efforts have largely been ad hoc and ineffective (although this is beginning to change). There is considerable demand for scientific assessment from multilateral environmental agreements such as the UNFCCC, from national governments that are faced with both environmental commitments and new environmental policies, from the private sector (which is increasingly monitored and regulated by the insurance industry), and from regional organizations. However, there remains a lack of reliable scientific information and assessment in a variety of areas, most notably in relation to the human dimensions of environmental change. We may argue about the availability and reliability of data on soil type and quality, but there is little argument about the dearth of information on property rights and land tenure systems, or what measures of sustainability are applicable in communities in Africa. There is also a crucial need to better understand the adverse impacts of the new array of multilateral environmental agreements such as the Kyoto Protocol. Not only are governments unwilling to address this issue, but few assessment bodies consider the long term. Recent negotiations around Kyoto have resulted in the establishment of a development fund to assist countries that may be adversely impacted by the agreement, but the amounts of money are relatively minor. The only way to meet this challenge is through integrated assessment models that include the broad participation of all the stakeholders involved. Such models would link biophysical and socio-economic considerations of climate change into a coherent decision-making structure. Although the first two chapters of this volume deal separately with scientific and human challenges, they are, in fact, closely linked. Until we develop models that help us understand how institutions affect climate, which in turn affects economies and societies, we will lack the ability to make credible decisions about how best to deal with climate change.

Conclusion

What are the key human challenges of climate change? This chapter has identified a number of important issues, but three stand out. First, we must ensure that the consequences of climate change do not overwhelm the human adaptive capacities in various regions. We need to assess socio-economic vulnerability to climate change and promote adaptive mechanisms so that impacts—particularly of extreme climate events—will be minimized. Since the greatest human impact of climate change is expected to be on the poor and disenfranchised, we must focus more attention on these groups. Second, we must develop appropriate institutions to ensure that both mitigation and adaptation strategies are adequately understood and implemented. The UNFCCC process is evolving along these lines and it should be continued. Canada can play a major role in the continued development of the UNFCCC and the eventual implementation of any global agreement. Canada should thus work toward meeting its Kyoto commitments. And third, we must ensure that equity is a key component of all discussions and agreements regarding climate change. This means that developed countries must own up to their responsibility in contributing most of the measured increase in greenhouse gas levels and to the fact that they have the greatest ability to pay for stabilizing global emissions. The challenge of ensuring that equity is central to all discussions is undoubtedly the most important human challenge of climate change, and likely the most difficult to carry out.

References

Agarwal, A., S. Narain, and A. Sharma. (1999). *Green Politics.* New Delhi: Centre for Science and the Environment.

Allen, J., and D. Barnes. (1985). The Causes of Deforestation in Developing Countries. *Annals of the Association of American Geographers* 75(2): 163–84.

Brklacich, M., C. Bryant, B. Veenhof, and A. Beauchesne. (1998). Implications of Global Climatic Change for Canadian Agriculture: A Review and Appraisal of Research from 1984 to 1997. In *Canada Country Study: Climate Impacts and Adaptation,* ed. G. Koshida and W. Avis, 219–56. Downsview, ON: Environment Canada.

Claussen, E., and L. McNeilly. (1998). *Equity and Global Climate Change.* Washington, DC: Pew Center.

Darmstadter, J., P. Teitelbaum, and J. Polach. (1971). *Energy in the World Economy.* Baltimore: Johns Hopkins Press.

Eyles, J., and R. Sharma. (2001). *Infectious Diseases and Global Change: Threats to Human Health and Security.* Aviso Report No. 8. Victoria, BC: Global Environmental Change and Human Security Project, University of Victoria.

Falkenmark, M., J. Lundqvist, and C.G. Widstrand. (1989). Macro-Scale Water Scarcity Requires Micro-Scale Approaches. *Natural Resources Forum* 13(4): 258–67.

Gleick, P. (1996). Basic Water Requirements for Human Activities: Meeting Basic Needs. *Water International* 21(2): 83–92.

Homer-Dixon, T. (1991). On the Threshold: Environmental Changes as Causes of Acute Conflict. *International Security* 19(1): 5–40.

IPCC. (1995). *Equity and Social Considerations Related to Climate Change.* Nairobi, Kenya: United Nations Environment Programme.

———. (1996). *Climate Change 1995: The Consequences of Climate Change: Contribution of Working Group II to the Second Assessment Report of the Intergovernmental Panel on Climate Change.* New York: Cambridge University Press.

———. (2001). *Climate Change 2001: Impacts, Adaptation and Vulnerability.* Contribution of Working Group II to the Third Scientific Assessment Report of the Intergovernmental Panel on Climate Change. New York: Cambridge University Press.

Koshida, G., and W. Avis, eds. (1998). *Canada Country Study: Climate Impacts and Adaptation.* Executive Summary. Downsview, ON: Environment Canada.

Lonergan, S.C. (1993). Impoverishment, Population and Environmental Degradation: Assessing the Relationships. *Environmental Conservation* 20(4): 328–34.

———. (1999). *The Global Environmental Change and Human Security Science Plan.* IHDP Report No. 11. Bonn, Ger.: International Human Dimensions Programme on Global Environmental Change.

Lonergan, S.C., and D. Brooks. (1994). *Watershed: The Role of Freshwater in the Israeli-Palestinian Conflict.* Ottawa: IDRC Books.

Lonergan, S.C., R. Difrencesco, and M.-K. Woo. (1993). Climate Change and Transportation in Northern Canada: An Integrated Impact Assessment. *Climatic Change* 24(4): 331–52.

LUCC. (1998). *Land Use, Land Cover Change* (LUCC) *Implementation Strategy.* IHDP Research Report No. 10. Bonn, Germany: International Human Dimensions Programme on Global Environmental Change.

McMichael, A., A. Haines, R. Sloof, and S. Kovats, eds. (1996). *Climate Change and Human Health.* WH/EHG/96.7. Geneva: World Health Organization.

National Research Council. (2002). *Abrupt Climate Change: Inevitable Surprises.* Washington, DC: National Academy Press.

Ohlsson, L. (1999). *Environment, Scarcity and Conflict: A Study of Malthusian Concerns.* PhD dissertation. Göteborg: Department of Peace and Development Research, Göteborg University.

ORNL (Oak Ridge National Lab). (2001). *Trends Online: A Compendium of Data on Global Change.* Oak Ridge, TN: Carbon Dioxide Information Analysis Center. <http://cdiac.esd.ornl.gov/trends/trends.htm>.

Richards, J. (1990). Land Transformation. In Turner et al., 1990, 163–78.

Royal Society of Canada (2001). *Implications for Canada of the IPCC Assessment Reports: An Overview.* Ottawa: Royal Society.

Turner, B.L. et al. (1990). *The Earth as Transformed by Human Action: Global and Regional Changes in the Biosphere Over the Past 300 Years.* Cambridge, UK: Cambridge University Press.

UNDP. (1998). *Human Development Report.* New York: Oxford University Press.
———, (2000). *Human Development Report.* New York: Oxford University Press.
US National Research Council. (1999). *Global Environmental Change: Research Pathways for the Next Decade.* Washington, DC: National Academy Press.
Westing, A. (1986). *Global Resources and International Conflict: Environmental Factors in Strategic Policy and Action.* Oxford: Oxford University Press.
World Resources Institute. (1998). *World Resources, 1998–1999.* Washington, DC: World Resources Institute.
Young, O. (1999). *Science Plan: The Institutional Dimensions of Global Environmental Change.* IHDP Report No. 12. Bonn: International Human Dimensions Programme on Global Environmental Change.

4

Impacts of Climate Change in Canada

James P. Bruce and Stewart J. Cohen*

As GLOBAL EMISSIONS OF GREENHOUSE GASES continue to soar, the world's climate is changing ever more rapidly. Canada, being at northern latitudes where changes are expected to be greatest, has begun to feel adverse effects. Table 4.1 gives information about the changes affecting Canada to the end of 2000 and the projected changes into the present century. The latter are based mainly on the outputs of the Canadian Climate Model—an Atmosphere-Ocean General Circulation Model (Canadian AOGCM). This model has predicted reasonably well the changes that have occurred in Canada to date (Boer et al., 1998). The Canadian GCM and most others include the effect of particles or aerosols in the air, which partly offset the greenhouse gas effect. In these model outputs, it is assumed that atmospheric greenhouse gas concentrations will continue to increase due to continued global increases in emissions (primarily from burning of fossil fuels). In a few cases in Table 4.1, the consensus projections of the Intergovernmental Panel on Climate Change (IPCC)[1] 2001 report are cited instead of Canadian model outputs.

It must be recognized, however, that over the twenty-first century, the increase in emissions of greenhouse gases could have a wide range. If global economies, populations, and energy consumption from fossil fuels continue to increase at a rapid rate, carbon dioxide (CO_2) emissions could be more than triple pre-industrial levels by 2100. However, if energy conservation and use of renewable energies increase, and reductions in rates of economic and population growth occur, CO_2 emissions might increase only 25% by 2050. Bringing the Kyoto Protocol into force could start the

* Opinions expressed in this chapter are those of the authors, and do not necessarily reflect the views of the Government of Canada.

Table 4.1 Climate Change Projections and Observations for Canada

	Projected	**Observed to Date (2000)**
Global Mean Temperature	1.4–5.8°C (1990–2100)	+0.6 or −0.2°C (20th century)
Canadian Mean Temperature	2–4°C (CGCM: 1975–95 to 2040–60)	>1°C (20th century)
Total Precipitation	(2040–2060) 0 to 20% more in north slightly less in mid continent in summer ($HadCM_3$)	1950–1998 ++ at high altitudes, + at mid latitudes Southern Prairies little change
Streamflow (or soil moisture) Mid-continent	30% by 2050 $2 \times CO_2$ (CGCM)	−10% Southern Prairies (1967–1996)
Date of Spring Breakup	Earlier	Earlier: 82% of basins (1967–1996)
Extreme Rainfall	Years between heavy rain events reduced by half (2090)	Up to 20% increase in heavy 1-day falls in US and SE Canada (early summer)
Water Vapour in Troposphere (lower atmosphere)	Increase	Statistically significant increase over N. America except NE Canada
Mean Sea Level Rise	40–50 cm (mean IPCC projections) 1990–2100	10–20 cm (1900–1999)
Arctic Sea Ice Extent	−21 to −27% by 2050	−3% per decade since 1978 (year round ice extent)
Snow Cover Extent Dec., Jan., Feb.	−15% by 2050 N. America (CGCM)	−10% (1972–2000) Northern Hemisphere
Late Season Snow Pack in Rockies—Apr. 1	Less (more melt over winter)	30% less since 1976 Fraser River Basin
Glacier Retreat South of 60° N e.g., Glacier National Park	None left in Park (by 2030)	2/3 reduction in numbers in Park (from 150 to 50) (1850–1990s)
Severe Winter Storms Frequency and Intensity	15% to 20% increase $2 \times CO_2$ (CGCM)	(1959–1997) • N of 60°N Increased frequency and intensity • S of 60°N Increased intensity

Source: Data from Akinremi et al., 1999; Angel and Isard, 1998; Boer et al., 2000; Carnell and Senior, 1998; Gregory et al., 1997; IPCC, 2001; Karl et al., 1995; Lambert 1995; McCabe et al., 2001; Mekis and Hogg, 1999; Moore, 1996; Ross and Elliot, 1996; Sarnko et al., 2002; Stone et al., 2000; Zhang et al., 2000; Kharin and Zwiers, 2000.

Notes: $HadCM_3$ = Hadley Centre (UK) Climate Model version 3
$2 \times CO_2$ = doubled pre-industrial level of CO_2 equivalent (by latter half of 21st century)
CGCM = Canadian Global Climate Model (CCCMA) (Environment Canada, University of Victoria)
++ = significantly more
+ = more

world on this latter path. Uncertainty about future emissions is the largest source of uncertainty about climatic futures. Nevertheless, even with depressed world economies and a huge change in energy consumption patterns, atmospheric concentrations and rates of change in climate will continue to increase at least until mid-twenty-first century. This increase is inevitable because greenhouse gases, once emitted, pervade the whole global atmosphere and typically stay in the climate system for a century or more (except methane, which has a lifetime of only 10–15 years) (IPCC, 2001).

If little change occurs in current growth rates and fossil-fuel consumption, global atmospheric concentrations could rise to three or four times pre-industrial levels of CO_2, and some climate change effects could be catastrophic. Present elevated levels of atmospheric CO_2 are the highest that the Earth has experienced in over 400,000 years, and these conditions could result in nasty surprises and sudden shifts in climate. The IPCC Reports include analyses of recent trends and projections by climate system models, and all indicate that changes over continental areas have been and will be greater than over oceans, and that changes will be greater than the global average at higher latitudes in the northern hemisphere (i.e., towards the Arctic).

Thus for Canada, a northern, largely continental country, changes will be greater than global average trends, and Arctic and Subarctic Canada will experience the greatest temperature and precipitation changes.

It is beyond the scope of this chapter to outline all of the climate change impacts being felt now and projected for the future across Canada. However, to give the reader a sense of the changes occurring and their impacts, a summary is given for the Arctic and Subarctic, for the southern Prairie provinces, and for the Great Lakes–St. Lawrence Basin in southern Ontario and Quebec (see Bruce, 2002).

The North

Sea ice

Changes of a profound nature are already occurring in the Western Arctic and Subarctic resulting from an average 1.5°C warming to date and a further warming of 5–7°C projected this century. These changes include the melting of sea ice, especially in the Beaufort Sea (Sarnko et al., 2002), thawing of permafrost (Cohen, 1997a, b), rising seas, shortening of the ice-cover season on rivers and lakes, and reduced ice-jam flooding in deltas and productive river valley lands. These changes affect wildlife such as polar bears and seals dependent on the ice regime, and human communities dependent on hunting over ice-covered seas (Ashford and Castledon, 2001). The loss of sea ice means that larger waves impinge on the northwestern coast where permafrost thawing softens the shores, and rising sea levels compound the problems of coastal erosion affecting communities such as Tuktoyaktuk (Shaw et al., 1998).

Permafrost

Problems are developing in the maintenance of buildings, utilities, pipelines, roads, and railroads in the face of land slumps from the thawing of ice-rich permafrost. The greatest immediate risks are where permafrost is present in areas where air temperatures currently average higher than −2° C, for example in much of the Mackenzie River Basin south of Great Bear Lake. Both thawing of permafrost and coastal erosion require remedial action for existing infrastructure, and these factors must be taken into account in future building and planning (Cohen, 1997a, b).

Vegetation

Much of the Arctic has recently experienced warmer temperatures in winter and spring, resulting in a lengthening of the growing season. Future warming would continue this trend, and species are likely to shift to higher latitudes and elevations. Forest fire severity ratings are expected to increase with higher temperatures in southern Yukon and the Northwest Territories, signalling more frequent and extensive fires and a lengthening of the fire season. Impacts on wildlife (such as caribou) are more difficult to assess, but would likely be unfavourable for those that prefer cold climates (IPCC, 2001).

Transportation

Transportation is being affected in the Mackenzie Basin and other areas where northern community resupply depends upon trucks using winter ice roads. The period of reliable ice road travel has been shortening with the warming climate, reducing the usefulness of this transport mode. At the same time, the period of high enough water levels and flows on the Mackenzie River to carry extensive barge traffic is also getting shorter, with more frequent low flow periods in late summer and autumn (Cohen, 1997a). These observed trends will accelerate with further warming. Shipping through the Beaufort Sea and the Northwest Passage will become much easier with melting ice cover (see Human Adaptation below).

Nunavut: Eastern Arctic

It should, at the same time, be noted that little or no warming has occurred in much of eastern Nunavut or elsewhere in northeastern Canada. This is consistent with atmosphere-ocean climate model results, which show much less rapid warming or even cooling in future in northern Quebec, Labrador, Baffin Island and Baffin Straits due to changes in ocean circulation and ice patterns (Boer et al., 1998).

Toxic contaminants

An environmental issue that affects all of the Canadian Arctic and Subarctic is contamination by toxic chemicals. For the most part, these originate in the industrial regions of the northern hemisphere and are transported, mainly by the atmosphere, onto Arctic waters and land. Many such contaminants, both toxic metals and organic compounds, are first deposited in water bodies to the south, such as the Great Lakes. In warm water conditions, they re-volatilize[2] to the atmosphere to be transported further until they reach regions where the water remains cold all year. There, in the Arctic and Subarctic, they bioaccumulate and become a hazard to the health of creatures at the top of the food chain, including humans. Long-lived organochlorines, like PCBs and DDT, persist in the Arctic long after their use was banned in many countries. Coal-burning power plants and municipal waste incinerators are substantial sources of Arctic mercury, which also reaches unhealthy levels in blood of some indigenous people. There is little research on the impact climate change will have on the transport of these contaminants into the Arctic. However, several scientists have pointed out that, as southern water temperatures rise to the point where volatilization occurs more frequently, these southern waters will tend to cleanse themselves more rapidly at the expense of greater deposition in the still relatively cold waters of the North.

Storms

Throughout the area north of 60°N severe winter storms have been increasing in both frequency and intensity over the past three to four decades putting hunters and travellers far from home base at ever more serious risk (McCabe et al., 2001). This trend is projected to continue (Carnell and Senior, 1998).

Human adaptation

What all this means for Canada's northern communities is that the North will be faced with additional climate-related challenges superimposed on rapid changes in northern economies and institutions (Cohen, 1997a, b; Ashford & Castledon, 2001). In the last decade, we have seen the creation of Nunavut in the eastern Arctic, the growth of tourism, and the onset of diamond mining. There are growing pressures to expand oil and gas exploration. Expanded wage-based activities may not be as vulnerable to climate change as traditional harvesting, although ice and permafrost instability will lead to problems for buildings, winter roads, and perhaps mines. Traditional indigenous lifestyles may also be disrupted by these changes, and national sovereignty will be an increased concern as Arctic waters become more accessible.

Dialogue on climate change is expanding in this region as communities and governments consider their development goals and various choices for meeting them. The West Kitikmeot Slave Study (WKSS) on the regional effects of expanded mining north of Great Slave Lake includes observations of the ways recent warming has been changing ice conditions, snowmelt, and vegetation patterns (WKSS, 2001). The community of Sachs Harbour on Banks Island has also observed these changes (Ashford and Castledon, 2001). However, the impact of these and future changes will vary regionally, and there is a belief that any damages or benefits that may accrue from climate change will be influenced by development and lifestyle choices (Table 4.2). The Manitoba Climate Change Task Force's inclusion of Northern issues in its report (2001), the establishment of the Northern Climate Exchange in Whitehorse Yukon (NCE, 2001), inclusion of Northern concerns in emergency preparedness planning within the Quebec government (Jaimet, 2002), and the initiation of the Arctic Climate Impact Assessment by the Arctic Council (2003), illustrate the growing interest in developing a Northern capability to learn more about climate change impacts and potential responses.

The Inuit, Dene, and Métis have shown themselves to be very resilient to fluctuations in climate and resource availability, but many are becoming dismayed by the longer-term changes they are experiencing and by the physical and social impacts on their hunting and fishing-based communities. Successful adaptation will require these communities to go beyond the experience of their elders and their practical knowledge of the land, sea, and ecological systems (Ashford and Castledon, 2001). As temperatures warm, exploitation of northern resources of oil, gas, and minerals, and potentially much greater use of Arctic waters for transportation, will become increasingly viable. These activities must be carried out in a manner sensitive to protection of the existing environment and communities of the North.

Table 4.2 Opinions of Residents of Aklavik Regarding Future Regional Impacts of Climatic Change within Different Visions of Lifestyles

Impact	Continued Reliance on Subsistence Activities	Greater Reliance on Wage Economy & Economic Development
Greater flooding	–	–
Muddy road conditions	?	–
Insulation of buildings	+	+
Easier water delivery	?	+
Less time spent waiting out cold conditions	+	+
Outdoor meat storage	–	–
Uncomfortably hot in summer	–	–
Increased summer insects	–	–
Shorter winter road season	?	?
Longer water shipping season	?	+
Mode of transport	?	?
Infrastructure of camps	–	?
Location of camps	–	?
Changes in wildlife habitat	–	?
Increased sediment loading	–	–
Thinner ice	–	–
Greater snowfall	?	–
Variability in timing & consistency of break-up & freeze-up	–	–
Longer ice free season	?	+
Shoreline erosion & lowland flooding	–	–
Greater variability in decisions/perceptions	–	?

Source: Aharonian, 1994, p. 419.

Notes: + =positive impact; – = negative impact; ? = impact is indeterminate. See text referring to Table 4.2 on page 78.

Southern Prairie Provinces

Water

The driest region in Canada is the southern Prairies, and evidence indicates this region is drying further with the changing climate. Both average annual and annual minimum stream flows have been declining on most rivers over the past 30 years (Zhang et al., 2001). Models project a continued drying trend, with little or no change in precipitation and significant evaporation increases with higher temperatures (Wetherald and Manabe, 1995). In the summer growing season, rainfall is also most likely to be reduced slightly in the future as it has been in recent years (Gregory et al., 1997).

Glacier-fed streams are at greatest risk. In general, flows first increase as glaciers begin to melt, but after a period they decrease as the volume of ice available for melt shrinks. Recent studies of paired watersheds on the eastern slopes of the Rockies in the North Saskatchewan Basin, show more rapid decline in flows in the glacier-fed rivers than in the non-glaciated basins (Pietronio, 2001). This suggests that for this critical basin, and probably for the South Saskatchewan system as well, glacier contributions have already peaked, and progressively less water will be available to the southern Prairies from glacier sources. With both of these downward pressures on water availability and increasing demand from communities and agriculture, water allocation and conservation issues will be increasingly critical (Bruce et al., 2000). With lower flows in the Saskatchewan-Nelson system, hydropower production will decline unless further storage and production facilities are installed.

Climate models suggest that rainfall in future will occur more frequently in severe storms or heavy bursts punctuating dry periods (Kharin and Zwiers, 2000). Poor water quality, in Prairie surface waters, tends to occur under two conditions:

- when flows are low and dilution of wastes is reduced; and
- when flows are abnormally high due to runoff from agricultural lands and urban areas, contributing pathogens and chemicals from animal wastes, agricultural areas, and urban lands.

For smaller watercourses on the Prairies, both conditions are likely to be more frequent and exacerbated by trends towards more intensive farm animal operations. In addition, some pathogens such as cryptosporidium and algal blooms, thrive in the warmer waters projected with climate change. Increasing vigilance is needed to protect water supplies (Bruce et al., 2000).

Agriculture

Higher temperatures have mixed effects on various crops, but where reductions in yield due to heat stress occur, they may be offset by the stimulation of increased CO_2 in the atmosphere (IPCC, 2001). A range of studies suggests a slight decline worldwide in cereal production due to these factors, but in North America this could probably be offset by suitable adaptation measures. A longer growing season provides more opportunities for adaptation (IPCC, 2001). However, on the Prairies the availability of water and soil moisture will be a far more important factor in crop production. Most climate projections and evidence of the past few decades indicate significant drying: 30% less soil moisture on average by 2050 in the Canadian model (Boer et al., 1998). These projections suggest more frequent drought with crop failures or reduced yields. Efforts to find and use more drought tolerant crops will be important. With projections of more intense rainfall events, greater use of measures to prevent soil erosion is also needed. Wind erosion will also increase in drought periods. Irrigation water requirements will be increasingly difficult to meet, especially in the dry southwest with declining flows and groundwater levels. Diversions from the North, greater conservation efforts, closer monitoring of shared water agreements between the US and Canada and between provinces, abandonment of very dry areas, reduction of city demands—all of these may have to be considered, along with their political, environmental, and economic consequences (Bruce et al., 2000).

Urban areas

Problems to be faced by urban areas include growing demands for increasingly scarce water supplies of good quality, increased health problems with heat stress and longer smog episodes (but fewer deaths due to severe cold), and drainage problems with more intense rainfalls (see Great Lakes-St. Lawrence basin) (Bruce et al., 1999).

Southern Ontario and Southwest Quebec: Great Lakes-St. Lawrence Basin

Changes in climate

In this industrialized and heavily populated region, mean temperatures to date have risen by about 0.6° C and are expected to continue to rise by another 2–3° C by 2050. Models give conflicting predictions about whether annual precipitation will increase slightly or decline, so a hypothesis of little future change in total precipitation might be assumed, with more of the precipitation falling as rain and less as snow. Under most models, evaporation losses are expected to more than offset any changes in precipitation. Water issues, including pollution problems and water-levels of the Great Lakes will require close monitoring, but health issues in and near urban areas, flash floods and drainage design, agriculture, and recreational impacts are also of significant concern as the climate changes (Bruce et al., 2000).

Great Lakes–St. Lawrence system

The extent of climate change impacts on the levels and flows of the Great Lakes remains a subject of disagreement among scientists. While most models suggest little precipitation change and increasing evaporation losses, especially from the large Upper Lakes (Superior–Huron) at least one (Hadley Centre 2)[3] predicts considerably more precipitation over the basin (although the Hadley Centre 3 model results are drier and close to other model outcomes). Thus, the majority of models indicate a significant lowering of lake levels (Mortsch et al., 2000). For the Canadian model, by 2050 these drops would be 0.3 m for Lake Superior, 1 m for Huron, 0.8 m for Erie, and 0.5 m for Ontario. These declines would result in a loss of 1.3 m water depth in Montreal harbour on the St. Lawrence. A drop in harbour water levels of only 30 cm in the 1988–91 period resulted in a 15% reduction in tonnage handled.

Although levels are still slightly above long-term record lows, recently experienced falls in lake levels (except on Lake Ontario), indicate the impacts of continuing lower levels and flows. Commercial shipping would be hampered, not only in Montreal harbour. For a 2.5 cm lowering of Lakes Michigan–Huron a cargo ship must reduce loads by 90–115 tonnes, a cargo reduction of some US$25,000 per trip. Hydroelectric production at Niagara and on the St. Lawrence would be reduced; losses were 19% and 26% of production respectively in the 1960s low-water period. Dredging channels would be costly and would stir up contaminated sediments. However, shore property owners would inherit more land: a 1.6 m water level decline on shallow Lake St. Clair would displace the shoreline by 1–6 km (Mortsch et al., 2000; Mortsch, 1998).

The ice regime of the Great Lakes is being affected by warming with projected changes in ice cover duration. This is expected to be reduced by 8 to 13 weeks by the second half of this century on Lake Erie, as an example. This will lengthen the shipping season, but in some regions lake-effect snows will be prolonged and intense, especially east of Lake Huron, Georgian Bay, and Lake Erie (mostly in the Buffalo–Fort Erie area). The open waters will also exacerbate lake level decline since evaporation proceeds for a longer period (Sousounis and Bisanz, 2000). A recent episode of warm weather illustrated some of these impacts, including a longer ice-free season and reduced cargo limits for shipping (Assel et al., 2000). Studies have also suggested that shoreline wetlands, groundwater supplies, fisheries, agriculture, and hydroelectric production would be adversely affected (Mortsch, 1998; Sousounis and Bisanz, 2000).

In the middle reaches of the St. Lawrence River, higher sea levels and lower freshwater flows, on average, are expected to result in increased salinity upstream, affecting aquatic ecosystems and water supplies.

Groundwater and tributary water supplies

With warmer summers, water demand from groundwater and tributary rivers like the Grand and Ottawa are increasing for both consumptive and recreational uses. Trends to date in streamflow and water levels have been mixed with some streams showing declines in low flows and others small increases. Groundwater provides 50% of Great Lakes tributary streamflow in Ontario, and 90% of rural residents use groundwater for domestic purposes. The Canadian GCM predicts a 19% drop in groundwater levels and their contribution of base flow or minimum flows to streams by 2050 (Piggott, 2001).

Implications for water management

Coping with changes to surface water and groundwater will be a Canada-US challenge (Browne et al., 1997). There have been many examples of cross-border cooperation in research and dialogue in this region, including those organized through the International Joint Commission. However, while there have been suggestions for developing a proactive strategy for adapting to climate change, including much discussion about water diversions, no clear consensus has emerged, other than to step up efforts to monitor water resource supplies and use, and to develop a basin management program (IJC, 2000). This lack of consensus suggests the need for a science-based water allocation system for more comprehensive basin-wide management programs such as that now in place in the Grand River basin.

Flash flood potential and drainage design

As noted above, heavy rainfalls of short duration (1 day or less) are projected to be more frequent in warmer conditions. The Canadian model for doubled CO_2 predicts that the average interval between heavy rain events will be reduced by 50%, e.g., a heavy rain equalled or exceeded only once in 20 years over a long period of time (a 20-year return period rain) becomes a 10-year return period event (Kharin & Zwiers, 2000). In the Great Lakes-St. Lawrence Basin and Atlantic Canada, high rain intensity events in spring and early summer have shown significant increases from 1950 to 1995 (Stone et al., 2000). A recent study indicates that a 6–7% increase in urban and suburban runoff volumes occurs for every 5% increase in rain intensity. This poses a dilemma for municipalities—do they accept an increased frequency in the overflow and flooding of storm sewers, or do they enlarge sewer capacity at substantial cost? Storm-water storage in combined storm and sanitary sewer systems would have to be increased by 11–16% for a 5% increase in rainfall intensity in a 100-year storm in order to prevent overflows and discharge of sanitary wastes.

Agriculture

Increased frequency of heavy rains also increases episodes of surface runoff from agricultural lands, with potential for, and tragic experience with, farm chemicals and E. coli from animal waste entering rivers and groundwater supplies. Soil erosion is expected to increase, but, in general, longer growing seasons should benefit most agricultural production in the Great Lakes-St. Lawrence Region (Bruce et al., 2000).

Urban air pollution

Smog episodes will be longer and more intense in this region. An estimated 1900 premature deaths per year occur at present in Ontario from smog and air pollution, with additional cases in the Montreal area and southern Quebec. More intense and prolonged heat waves will make this an even more serious public health issue, when both air pollution and high heat stress affect vulnerable populations of asthmatics and the elderly. Remedial actions are urgent. Reducing dependency on fossil fuels can reduce local air pollutants and contributions to greenhouse gas forcing of climate at the same time.

Recreation

Winter snow-based recreation in southern Quebec and southwest Ontario will have shorter seasons in future, although snow amounts could increase at ski resorts to the east of Lake Huron and Georgian Bay with a longer period of lake-effect storms. A longer summer recreation season is expected.

Winter storms

With a trend towards more intense winter storms, paralyzing snow and freezing rain events are expected to be more frequent (Lambert, 1995).

Conclusions

Our society is currently well adapted, by both where and how we live and work, to the climate of the *past* century. While there are some potentially positive effects in Canada as these climatic conditions change, many more severe and adverse impacts will be increasingly felt in coming decades. Some can be addressed by adaptation measures, but these measures can be costly (see chap. 8). Other impacts, especially on ecosystems, cannot be readily affected by human adaptation measures.

Notes

1 The Intergovernmental Panel on Climate Change (IPCC) was established by UN agencies in 1988 to assess the results of available scientific knowledge of climate change. It includes natural and social scientists from most UN countries and from universities, government and industry. Major reports were issued in 1990, 1995 and 2001. Typically 1000–2000 scientists participate as authors and reviewers.

2 An evaporation process that releases the substance in a gaseous form to the atmosphere. When this occurs several times, with the molecules moving gradually further north from water body to water body, it is often called the "grasshopper effect."

3 The Hadley Centre was established by the UK and well endowed to become one of the world's leading climate change modelling and climate studies institutes.

References

Abraham, J., T. Canavan, and R. Shaw, eds. (1997). *Climate Change and Climate Variability in Atlantic Canada. Vol. 4, Canada Country Study: Climate Impacts and Adaptation.* Bedford, NS: Environment Canada.

Aharonian, D. (1994). Land Use and Climate Change: An Assessment of Climate-Society Interactions in Aklavik, NWT. In *Mackenzie Basin Impact Study* (MBIS) *Interim Report #2*, ed. S.J. Cohen, 410–20. Downsview, ON: Environment Canada.

Akinremi, O.O., S.M. McGinn, and U.W. Cutworth. (1999). Precipitation Trends on the Canadian Prairies. *Journal Climate* 12(10): 2996–3003.

Angel, J.R. and S.A. Isard. (1998). The Frequency and Intensity of Great Lakes Cyclones. *Journal of Climate* 11(1): 61–71.

Arctic Council. (2003). Activities Listed on the Arctic Council web site, at <www .arctic-council.org>

Arora, V.K., and G.J. Boer. (2001). Effects of Simulated Climate Change on the Hydrology of Major Miver Basins. *Journal Geophysical Research* D4: 3335–48.

Ashford, G., and J. Castledon. (2001). *Inuit Observations on Climate Change: Final Report.* Winnipeg: International Institute for Sustainable Development (IISD) <www.iisd.org>.

Assel, R.A., J.E. Janowiak, D. Boyce, C. O'Connors, F.H. Quinn, and D.C. Norton. (2000). Laurentian Great Lakes Ice and Weather Conditions for the 1998 El Nino Winter. *Bulletin of the American Meteorological Society* 81(4): 703–17.

Boer, G.J., G. Flato, and D. Ramsden. (1998). A Transient Climate Change Simulation with Greenhouse Gas and Aerosol Forcing: Projected Climate for the Twenty-First Century. *Climate Dynamics* 16(6): 427–50.

Browne, J., L. Mortsch, S. Cohen, K. Miller, G. Larson, H. Gilbert, and L. Mearns. (1997). Vulnerability Assessment: US/Canadian Boundary Regions. In *Climate Variability and Transboundary Freshwater Resources in North America.* Montreal: NAFTA Commission for Environmental Cooperation.

Bruce, J.P. (2002). *Climate Change Effects on Regions of Canada.* Report for Federation of Canadian Municipalities, 23 pp. Ottawa: Global Change Strategies International.

Bruce, J.P., I. Burton, I.D.M. Egener, and J. Thelen. (1999). *Municipal Risks Assessment: Investigation of the Potential Municipal Impacts and Adaptation Measures Envisioned as a Result of Climate Change.* Ottawa: Global Change Strategies International. National Secretariat on Climate Change.

Bruce, J.P., I. Burton, H. Martin, B. Mills, and L. Mortsch. (2000). *Water Sector: Vulnerability and Adaptation to Climate Change.* Ottawa: Global Change Strategies International. <www.gcsi.ca/downloads/ccafwater.pdf>.

Carnell, R.E., and C.A. Senior. (1998). Changes in Mid-Latitude Variability Due to Greenhouse Gases and Sulphate Aerosols. *Climate Dynamics* 14: 369–83.

Cohen, S.J., ed. (1997). *Mackenzie Basin Impact Study* (MBIS) *Final Report.* Downsview, ON: Environment Canada.

———. (1997). What If and So What in Northwest Canada: Could Climate Change Make a Difference to the Future of the Mackenzie Basin? *Arctic* 50(4): 293–307.

Environment Canada. (1995). *The State of Canada's Climate: Monitoring Variability and Change.* A State of Environment Report. Ottawa: Environment Canada.

Gregory, J.M., J.F.B. Mitchell, and A.J. Brady. Summer. (1997). Drought in Northern Mid-Latitudes in Time-Dependant CO_2 Climate Experiment. *Journal of Climate* 10(4): 662–86.

Intergovernmental Panel on Climate Change (IPCC). (2001). *Reports of Working Groups I, II and III.* Cambridge, UK: Cambridge University Press.

Jaimet, K. (2002). Climate Change Threatens Far North. *Ottawa Citizen,* 12 April.

Karl, T.R., R.W. Knight, and N. Plummer. (1995). Trends in High Frequency Climate Variability in the Twentieth Century. *Nature* 377: 217–20.

Kharin, V.V. and F.W. Zwiers. (2000). Changes in Extremes in an Ensemble of Climate Simulations with a Coupled Atmosphere–Oceans GCM. *Journal of Climate* 13(21): 3760–88.

Lambert, S.J. (1995). The Effect of Enhanced Greenhouse Warming on Winter Cyclone Frequencies and Strengths. *Journal of Climate* 8: 1447–52.

Manitoba Climate Change Task Force (MCCTF). (2001). *Manitoba and Climate Change: Investing in Our Future.* Winnipeg: Manitoba Climate Change Task Force.

Marsh, P., and L.F.W. Lesack. (1997). Climate Change and the Hydrologic Regime of Lakes in the Mackenzie Delta. *Limnology, Oceanography* 41: 849–56.

McCabe, G.J., M.P. Clark, and M.C. Serreze. (2001). Trends in Northern Hemisphere Surface Cyclone Frequency and Intensity. *Journal of Climate* 14(12): 2765–68.

Mekis, E. and W. Hogg. (1999). Rehabilitation and Analysis of Canadian Daily Precipitation Time Series. *Atmosphere-Ocean* 37(1): 53–85.

Moore, R.D. (1996). Snowpack and Runoff Responses to Climatic Variability, Southern Coast Mountains. *British Columbia. Northwest Science* 40(4): 321–22.

Mortsch, L.D. (1998). Assessing the Impact of Climate Change on the Great Lakes Shoreline Wetlands. *Climatic Change* 40: 391–406.

Mortsch, L., H. Hengeveld, M. Lister, B. Lofgren, F. Quinn, M. Slivitzky, and L. Wenger. (2000). Climate Change Impacts on the Hydrology of the Great Lakes–St. Lawrence System. *Canadian Water Resources Journal* 25(20): 153–79.

Northern Climate Exchange (NCE). (2001). *Whitehorse Declaration on Northern Climate Change. Proceedings of Climate Change in the Circumpolar North: Summit and Sustainable Technology Exposition,* 19–21 March 2001, Whitehorse. <www.taiga.net/nce>.

Pietronio, A. (2001). *Impact of Climate Change on the Glaciers of the Canadian Rocky Mountain Slopes and Implications for Water Resources: Related Adaptation in the Canadian Prairie.* CCAF Impacts and Adaptation Project. Ottawa: Natural Resources Canada.

Piggott, A. (2001). *Groundwater and Climate Change Interaction in Southern Ontario.* CCAF Impacts and Adaptation Project. Ottawa: Natural Resources Canada.

Reeve, N., and R. Toumi. (1999). Lightning Activity as an Indicator of Climate Change. *Quarterly Journal Royal Meteorological Society* 125: 893–903.

Ross, R.J. and W.P. Elliot, (1996). Tropospheric Water Vapour Climatology and Trends over North America, 1973–93. *Journal of Climate* 9(12): 3561–74.

Sarnko, O.A., G.M. Flato, and A.J. Weaver. (2002). Improved Representation of Sea Ice Processes in Climate Models. *Atmosphere–Ocean* 40(1): 21–43.

Schindler, D. (1997). Widespread Effects of Climatic Warming on Freshwater Ecosystems in North America. *Hydrologic Processes* 11: 1043–67.

Schindler, D.W. et al., (1990). Effects of Climate Warming on Lakes of Central Boreal Forest. *Science* 250: 967–70.

Shaw, J., R.B. Taylor, D.L. Forbes, H.H. Ruz, and S. Solomon. (1998). *Sensitivity of the Coasts of Canada to Sea-Level Rise.* Geological Survey of Canada Bulletin 505. Ottawa: Natural Resources Canada.

Stone, D.A., A.J. Weaver, F.W. Zwiers. (June 2000). Trends in Canadian Precipitation Intensity. *Atmosphere–Ocean* 38(2): 321–47.

Sousounis, P.J. and J.M. Bisanz, eds. (2000). *Preparing for a Changing Climate—The Potential Consequences of Climate Variability and Change.* Great Lakes Overview. Report of the Great Lakes Regional Assessment Group for the US Global Change Research Program. Ann Arbor: University of Michigan.

West Kitikmeot Slave Study (WKSS). (2001). *West Kitikmeot/Slave Study Society Final Report, Yellowknife.* <www.wkss.nt.ca>.

Wetherald, R.T., and S. Manabe. (1995). The Mechanisms of Summer Dryness Induced by Greenhouse Warming. *Journal of Climate* 8(12): 3096-108.

Whitfield, P.H., and A.J. Cannon. (2000). Recent Variations in Climate and Hydrology in Canada. *Canadian Water Resources Journal* 25(1): 19-65.

Zhang, X., K.D. Harvey, W.D. Hogg, and T.R. Yuzyk. (2001). Trends in Canadian Streamflow, Water. *Resources Research* 37(4): 987-98.

Zhang, X., L.A. Vincent, W.D. Hogg, and A. Niitsoo. (2000). Temperature and Precipitation Trends in Canada during the 20th Century. *Atmosphere-Ocean* 36(3): 395-429.

Zwiers, F.W., and V.V. Kharin. (1998). Changes in the Extremes of the Climate Simulated by CCC GCM2 under CO_2 Doubling. *Journal of Climate* 11(9): 2200-22.

Part II

What Can We Do?

Terrestrial Carbon Sinks and Climate Change Mitigation

Nigel J. Livingston and G. Cornelis van Kooten

THIS CHAPTER ADDRESSES THE ROLE that terrestrial vegetation, particularly trees, might play in mitigating the effects of climate change. We specifically focus on the question of whether forests and agricultural land can be managed or manipulated to maximize carbon (C) uptake and so offset anthropogenic emissions of greenhouse gases. Photosynthesis is the most important biological mechanism by which plants take up carbon in both aquatic and terrestrial environments. This carbon is converted to sugars and subsequently to structural compounds such as cellulose and lignin. However, almost all of the "fixed" carbon is eventually returned to the atmosphere by the process of respiration. Thus, carbon sequestration by vegetation and soils can only, at best, be a temporary solution to increasing anthropogenic emissions. Nonetheless, it could be argued that appropriate use of our forests (and soils) will be critical in providing more time for the development of non-disruptive, long-term, or even permanent technological and socio-economic solutions to the problem of global warming.

Global Carbon Stocks and Land-Use Changes

As outlined in chapter 2, the atmospheric concentration of carbon dioxide (CO_2) has increased from about 280 parts per million by volume (ppmv) around 1800 to almost 370 ppmv today. This increase has happened because the rate at which CO_2 is being emitted from the burning of fossil fuels and deforestation (among other sources) exceeds the rate at which it is absorbed and stored on land and in oceans. Fossil-fuel emissions total approximately 6 Gt (gigatonnes, or billion tonnes) per year, and estimates suggest that terrestrial vegetation and soils take up only about 40% of global CO_2 emissions from human activities. As a result, carbon is accumulating in the atmosphere at an annual rate of about 3 Gt (fig. 5.1). However, there is considerable uncertainty and controversy surrounding estimates of global and regional carbon stocks and fluxes (rates of carbon exchange between the surface and the atmosphere). This is due to global differences in measurement methodologies and the difficulty in making reliable, consistent, and meaningful spatial and temporal measurements.

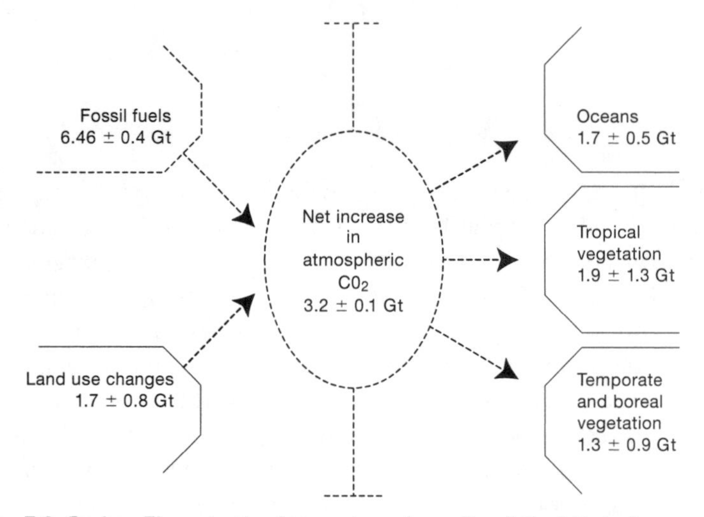

Figure 5.1. Carbon Flows to the Atmosphere from Fossil Fuel Emissions and Land Use Changes, and from the Atmosphere to Land and Ocean Sinks

Source: Royal Society, 2001.

The atmosphere contains approximately 760 gigatonnes of carbon (Gt C). This is about a third of the total amount of carbon held in vegetation (550 ± 100 Gt) and soils (1750 ± 250 Gt). Most of the carbon stored in terrestrial ecosystems is in forest soils and vegetation. For example, tropical and boreal forests (and soils) account for almost 40% of the carbon reservoir. By contrast, croplands account for only 5% of the total stored carbon. It takes approximately 10 years for all atmospheric CO_2 to exchange with land surfaces, but the net exchange is in disequilibrium because of natural climatic variability and the direct and indirect effects of human activity (including changes in land-use practices).

Carbon is taken up by plants through the process of photosynthesis—the largest-scale synthetic process on earth. Almost 50% of the total global photosynthesis is accomplished by marine organisms. However, when photosynthesis is expressed on a per unit area basis, the productivity of terrestrial ecosystems is much higher than that of marine ecosystems. The photosynthetic process (with the exception of photosynthetic bacteria) involves converting the conversion of light energy captured by pigments (the principal class of which are chlorophylls) into the chemical energy of organic molecules, primarily carbohydrates, using CO_2 and water from the environment: molecular oxygen is subsequently released. Energy is gained in the chemical bonds of the carbohydrates and is stored in high-energy compounds such as adenosine tri-phosphate (ATP). In the process of respiration, energy captured by photosynthesis is used to synthesize and maintain plant tissue, releasing CO_2 and consuming oxygen. The breakdown and utilization of plant tissue by microbial organisms in the soil also releases respiratory carbon.

Terrestrial carbon uptake by photosynthesis is estimated at 120 Gt per year, and plant respiration (carbon loss) is about half that value (fig. 5.2). The combined release of carbon through fires (4 Gt) and decomposition of soil organic matter (55 Gt) is also about 60 Gt per year. The net land carbon sink is about 3.2 Gt per year, which is rather small relative to the constituent

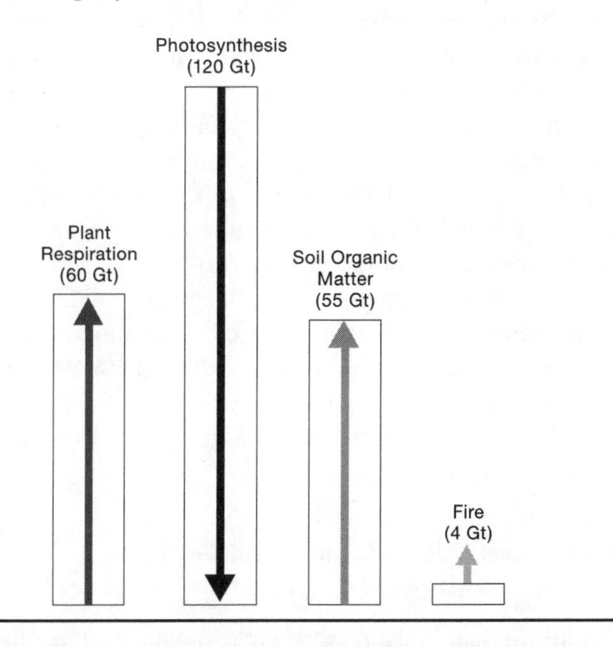

Figure 5.2. Estimates of Carbon Exchange in Gigatonnes of Carbon per Year between the Atmosphere and the Terrestrial Ecosystems

Arrows pointing away from the surface indicate a net loss of carbon to the atmosphere. *Source*: Royal Society, 2001.

fluxes. A relatively small change of about 10% in carbon uptake or loss from land surfaces would be almost equivalent to the entire (current) annual global use of fossil fuels.

Forests contain just under half of the global stock of carbon, so increasing or at least maintaining this carbon sink should be the main objective of strategies aimed at maximizing terrestrial carbon sinks. For example, it has been calculated by Moffat (1997) that, between 2008 and 2026, New Zealand could offset its carbon emissions from fossil-fuel burning by increasing forest plantations at a rate of 100,000 hectares (ha) per year. However, by 2045 all plantable land will be used up. Recent evidence suggests that in the northern hemisphere there has been a net increase in CO_2 uptake directly as a result of the reversion of agricultural lands to forests. Moffat (1997) estimated that, over the last 100 years, between 9 and 11 million hectares of formerly agricultural land in the southern and eastern US are now forested, and that, over the past 40 years, the increase in carbon stored in biomass and organic matter in forested land is equivalent to one quarter of all US greenhouse gas emissions over the same period. In temperate regions, it is estimated that land use and land cover changes in the 1990s resulted in an overall uptake of 0.8 Gt C per year (IPCC 2000).

By contrast, extensive deforestation in tropical regions has led to large losses of carbon to the atmosphere (about 1.6 Gt per year) and is considered to be the most important cause of carbon release arising from land-use change. This carbon is returned to the atmosphere through direct burning of vegetation and the breakdown of biomass and soil organic matter. Recent estimates indicate that the tropics as a whole are near equilibrium or a carbon/CO_2 source (Schimel et al., 2001). Again, there is considerable uncertainty in these estimates. It has been suggested, for example, that significant (and previously unaccounted for) amounts of carbon might be lost due to outgassing from rivers and wetlands in forested areas. Thus, in the Amazon Basin, up to 0.5 Gt C per year could be lost as a result of the respiration of organic matter transported from upland and flooded forests (Richey et al., 2002).

Land Use, Land-Use Change, and Forestry Mitigation Strategies

Forestry and agricultural management can reduce atmospheric CO_2 concentrations by either increasing terrestrial carbon storage (e.g., by planting more trees or by implementing agricultural practices that increase the incorporation of carbon into soil organic matter) or by decreasing biological emissions (e.g., by reducing deforestation or implementing soil conservation practices). The use of renewable biofuel (energy) crops as a substitute for fossil fuels also provides a means of decreasing CO_2 emissions. A recent comprehensive study undertaken by the Royal Society of the UK concluded that changes in agricultural and forestry practices and slowing deforestation have the potential to bring about a maximum of 25% of the reductions in CO_2 emissions by 2050 that are required globally to limit global warming. It also concluded that there is little potential for improvement thereafter. The IPCC (2001) estimates that, between 2000 and 2050, direct human management of terrestrial carbon sinks could sequester a maximum of 100 Gt C. About one third of this would be achieved through changes in agricultural management: some 14% from slowing deforestation and about 18% from replanting trees in the tropics. We discuss the forestry options and agricultural management in more detail below.

Forest Management and Afforestation, Reforestation, and Deforestation

Afforestation and reforestation lead to an increase in the amount of carbon stored on land. Afforestation refers to human activities that encourage growing trees on land that has not been forested in the past 50 years, while reforestation refers to human activities that encourage growing trees on land that was forested but has been converted to non-forest use. Deforestation, in contrast, leads to the release of carbon as a result of human conversion of forestland to non-forest use. Since most countries have not embarked on large-scale afforestation or reforestation projects in the past decade, harvesting trees during the five-year commitment period of the Kyoto Protocol (2008–2012) will cause a debit in the afforestation–reforestation–deforestation (ARD) account. Therefore, the Marrakech Accords permit countries, in the first commitment period only, to offset up to 9.0 megatons of carbon (Mt C) each year for the five years of the commitment period through (verified) forest management activities that enhance carbon uptake. If there is no ARD debit, then a country cannot claim the credit. Some countries are also able to claim carbon credits from business-as-usual forest management that need not be offset against ARD debits. Thus, Canada can claim credits of 12 Mt C per year, the Russian Federation 33 Mt C, Japan 13 Mt C, and other countries much lesser amounts—Germany 1.24 Mt C, Ukraine 1.11 Mt C, and remaining countries less than 1.0 Mt C. Japan expects to use forestry activities to meet 65% of its Kyoto Protocol obligation, while Canada can use forest management alone to achieve one-third of its emissions reduction target (but only 6% if projected 2010 emissions are used as the baseline from which to achieve the Kyoto target).

Surprisingly, despite the size of their forests and, in some cases, large areas of marginal agricultural land, there remains only limited room for forest-sector policies in the major wood-producing countries (Canada, Finland, Sweden, and Russia). We illustrate this using the Economic, Carbon and Biodiversity dynamic optimization model (TECAB) for northeastern British Columbia (Stennes, 2000; Krcmar and van Kooten, 2001). The model consists of tree-growth, agricultural and land-allocation components, and is used to examine the costs of carbon uptake in the grain belt–boreal forest transition zone of BC. Estimates of carbon uptake, extended to similar regions, provide a good indication of the costs of an afforestation–reforestation strategy for carbon uptake for Canada as a whole, and likely for other boreal regions as well. The study region consists of 1.2 million ha: nearly 10.5% constitute marginal agricultural land, and the remainder is boreal forest. The boreal forest is composed of spruce, pine, and aspen. For environmental reasons and to take into account other forest attributes (e.g., biodiversity and wildlife habitat needs), the area planted to hybrid poplar in the model is limited only to logged stands of aspen and marginal agricultural land. Other harvested stands are replanted to native species or left to regenerate on their own, depending on what is economically optimal. Carbon fluxes associated with forest management, wood product sinks, and so on are all taken into account. An infinite time horizon is employed, land conversion is not instantaneous (as assumed in some models), carbon fluxes associated with many forest management activities (but not control of fire, pests, and disease) are included, and account is taken of what happens to the wood after harvest, including their decay.

Study results indicate that upwards of 1.5 Mt C can be sequestered in the region at a cost of about $40/t or less. This amounts to an average of about 1.3 t ha^{-1}, or about 52 kg ha^{-1} per year over and above normal carbon uptake. If this result is applied to all of Canada's productive boreal forestland and surrounding marginal farmland, then Canada could potentially sequester some 10–15 Mt C annually by this means. This amounts to at most 7.5% of Canada's annual Kyoto-targeted reduction, well below the 22% that had been envisioned by other models (e.g., Canadian Pulp and Paper Association, 2000). This is a rather pessimistic conclusion given that plantation forests are generally considered a cost-effective means of sequestering carbon (Sedjo et al., 1995; Adams et al., 1999); boreal forests are globally marginal at best, and silvicultural investments simply do not pay, for the most part, even when carbon uptake is included as a benefit of forest management (van Kooten et al., 1993; Wilson et al., 1999). One option, therefore, is to plant only fast-growing species, such as hybrid poplar, on much of Canada's less valuable farmland.

There remains a great deal of uncertainty about planting hybrid poplar on a large scale, however, because it has not been done before. Drawbacks limit the viability of hybrid poplar:

- Relative to native species, hybrid poplar plantations have detrimental environmental impacts related to reduced biodiversity and susceptibility to disease (Callan, 1998).
- If transaction costs are associated with afforestation, this will increase carbon-uptake costs above what has thus far been estimated.
- Uncertainty about current and future stumpage values and prices of agricultural products makes landowners reluctant to convert agricultural land to forestry.
- Little is known about the potential of wood from afforested land as a biomass fuel.
- Recent research suggests that planting trees where none existed previously decreases the surface albedo (in the winter) and that this offsets any negative forcing expected from carbon uptake (Betts, 2000). Indeed, in some cases, a forestation program may even contribute to climate change rather than mitigating it as expected. This is more of a problem with coniferous than deciduous species, but it would not be entirely absent in hybrid poplar plantations.
- Leakage is a problem: large-scale afforestation is bound to lower wood-fibre prices, and existing woodlot owners elsewhere may reduce their forest holdings by converting land back to agriculture in anticipation. These economic adjustments are generally ignored in calculating the costs of individual afforestation or reforestation projects. Such leakages can be substantial, however, even as high as one-half of the carbon sequestered by the new plantations (Sohngen and Sedjo, 1999).

Other economic research indicates that, at least in western Canada, farmers in the zone where the greatest opportunities exist for converting agricultural land to forestry are also the most reluctant to do so because they (or their immediate forebears) are also the ones that have recently incurred the pains of clearing forestland for agriculture. They demand greater compensation for tree-planting than would otherwise be the case (van Kooten et al., 2002). Further, it might take upwards of 50 years to implement a tree-planting program of sufficiently large scale to make an impact on carbon uptake in Canada, if costs are to be kept reasonable so that tree planting is competitive with emissions reduction (van Kooten, 2000).

Now consider tropical deforestation. Tropical forests generally contain anywhere from 100 to 400 m^3 of timber per ha, although much of it may not be commercially useful. This implies that such forests store about 20–80 tonnes of carbon (t C) per ha in wood biomass, but this ignores other biomass and soil carbon. Estimates of total carbon stored in biomass for various tropical forest types and regions are given in Table 5.1. The carbon sink function of tropical soils is even more variable across tropical ecosystems (see Tables 5.2 and 5.3). This makes it difficult to make broad statements about carbon loss resulting from tropical deforestation. Certainly, there is a loss in carbon stored in biomass (which varies from 27 to 187 t C ha^{-1}). There may or may not be a significant loss in soil carbon depending on the new land use (agricultural activity) and the tropical zone. While conversion of forests to arable agriculture will lead to a loss of some 20–50% of soil carbon within 10 years, conversion to pasture may in fact increase soil carbon, at least in the humid tropics (Table 5.2). It is clear, however, that conversion of forest land to agriculture leads to a smaller carbon sink, with a greater proportion of the ecosystem's carbon stored in soils as opposed to biomass (Table 5.3). To address this problem, policies need to focus on the protection of tropical forests.

It may be difficult to prevent tropical deforestation, however, because the underlying or ultimate cause is government policy related to revenue and foreign exchange needs, and urbanization, and population control (Bromley 1999; van Kooten et al., 2000). Income and/or the foreign exchange generated from logging concessions (e.g., SE Asia) or new land use (cattle ranching in Brazil) are important for some governments in tropical regions. Governments may also permit or even encourage land-use changes as part of an overall policy to address urbanization pressure and general overpopulation in certain areas. Indonesia has moved peasants into outlying forested regions as a means of addressing over-crowding in Java, while Brazil has promoted development of the Amazon in order to encourage migration into the region and away from more urbanized areas to the South.

Other Land-Use Activities: Agriculture

In addition to ARD activities, certain forestland management and agricultural activities can lead to enhanced terrestrial carbon stocks. Thus, prevention of forest degradation (that does not lead to deforestation), revegetation (establishment of vegetation that does not meet the definitions of afforestation and reforestation), cropland management (greater use of conservation

Table 5.1 Carbon Content of Biomass, Various Tropical Forests and Regions

Country/Forest	Wet Tropical	Dry Tropical
Africa	187 t C ha^{-1}	63 t C ha^{-1}
Asia	160 t C ha^{-1}	27 t C ha^{-1}
Latin America	155 t C ha^{-1}	27 t C ha^{-1}

Source: Papadopol, 2000.

Table 5.2 Depletion of Soil Carbon Following Tropical Forest Conversion to Agriculture

Region	Soil C in Forest	New Land Use	Soil C Loss with New Land Use
Semi-arid	15–25 t C ha^{-1}	Shifting cultivation (arable agriculture)	30–50% loss in 6 years
Subhumid	49–65 t C ha^{-1}	Continuous cropping	19–33% loss in 5–10 years
Humid	60–165 t C ha^{-1}	Shifting cultivation pasture	40% loss in 5 years 60–140% of initial soil C

Source: Adapted from Paustian et al., 1997.

Table 5.3 Total Carbon in Tropical Ecosystems by Sink[a]

Land Use	Tree %	Understory %	Litter %	Root %	Soil %
Original forest	72	1	1	6	21
Managed & logged-over forest	72	2	1	4	21
Slash & burn croplands	3	7	16	3	71
Bush fallow	11	9	4	9	67
Tree fallow	42	1	2	10	44
Secondary forest	57	1	2	9	32
Pasture	<1	9	2	7	82
Agroforestry & tree plantations	49	6	2	7	36

Source: Woomer et al., 1999.

Note: [a]Average of Brazil, Indonesia, and Peru.

tillage, reduced summer fallow, and more set-asides), and grazing management (manipulation of the amount and type of vegetation and livestock produced) are activities that lead to enhanced soil organic carbon and/or more carbon stored in biomass. The problem with carbon stored by prevention of forest degradation and enhanced agricultural sinks is that these changes are likely ephemeral, and carbon flux is difficult to measure.

Estimates of the effects of both improved land-use management and changes in land use on net terrestrial carbon uptake are provided in Table 5.4. This table gives estimates of the potential of these activities for mitigating climate change; but it also demonstrates how current land uses have resulted in the release of C over time. For example, cultivation alone has

Table 5.4 Effects on Potential Net Carbon Storage of Land Use Activities, Excluding Afforestation and Reforestation[a]

Activity	Potential Area (10^6 ha)	Rate of C Gain (t C ha^{-1} yr^{-1})	Potential (Mt C yr^{-1}) 2010	2040
1. Improved Management of a Land Use	1,289	0.34	125	258
Cropland: reduced tillage, improved management of crop rotations, cover crops, etc.	(45.7%)		(60%)	(51%)
Rice paddies: better irrigation, improved residue management	153 (2.6%)	0.10	8 (<10%)	13 (<7%)
Agroforestry: better management of trees on cropland	400 (20.8%)	0.28	26 (46%)	45 (38%)
Grazing land: better management	3,401 (38.1%)	0.77	261 (36%)	523 (36%)
Forest land: enhanced silviculture, reduced degradation	4,051 (46.9%)	0.41	170 (59%)	703 (72%)
Urban land: tree planting, improved wood product management & waste management	100 (505)	0.30	2 (50%)	4 (50%)
2. Land Use Change				
Agroforestry: conversion of poor crop/grassland to agroforestry	630 (0%)	3.1	391 (0%)	586 (0%)
Restoring severely degraded land: to forest, crop or grass land	277 (4.3%)	0.25	4 (<20%)	8 (13%)
Grassland: converting cropland to grassland	1,457 (41.3%)	0.80	38 (63%)	82 (59%)
Wetland restoration: converting drained land back to wetland	230 (91.3%)	0.40	4 (100%)	14 (93%)
Global Total			**1,029 (39%)**	**2,236 (44%)**

Source: IPCC, 2000.

Note: [a]Contribution by developed countries provided in parentheses.

resulted in the historical release of 54 Gt C (Paustian et al., 1997). While a strategy to reduce forest degradation is addressed in Table 5.4, reforestation and afforestation programs are not considered.

The IPCC (2000) estimates that globally, changes in agriculture production have the potential to sequester between 22 and 44 Gt C between 2000 and 2050, about half that gained from improvements in forestry practices. This excludes the possible benefits of the use of biofuel crops.

Soils that are currently used for agriculture typically contain appreciably less organic carbon than equivalent soils under grasslands or forests because the removal of natural vegetation leads to both an accelerated breakdown of organic matter already in the soil and a decrease in input of organic carbon. In order to increase soil organic carbon, farmers would have to change their agronomic practices. In drier regions where tillage summer fallow is used to conserve soil moisture, this requires the use of chemical fallow or continuous cropping, or cessation of cropping altogether (i.e., return to grassland). In other agricultural regions, shifting from conventional tillage to reduced tillage or no tillage can increase soil organic carbon. Continuous cropping in place of summer fallow, reduced tillage, and zero tillage increase soil carbon by increasing plant biomass and/or reducing rates of decay of organic matter. Are such practices worth pursuing, and can they result in significant reductions in carbon flux?

West and Marland (2001) review previous studies comparing conventional tillage, reduced tillage, and zero tillage in terms of their carbon flux, and provide a detailed carbon accounting for each practice using US data. The researchers include not only carbon uptake in soils, but also carbon flux associated with machinery operations, production of agrochemicals, and so on. They conclude that reduced tillage does not differ significantly from conventional tillage in terms of carbon uptake benefits, and that zero tillage results in an average relative net carbon flux of −368 kg C ha^{-1} per year. Of this amount, −337 kg C ha^{-1} yr^{-1} is due to carbon sequestration in soil, −46 kg C ha^{-1} yr^{-1} due to a reduction in machinery operations, and +15 kg carbon ha^{-1} yr^{-1} due to higher carbon emissions from an increase in the use of agricultural inputs. While annual savings in carbon emissions of 31 kg C ha^{-1} yr^{-1} last indefinitely, accumulation of carbon in soil reaches equilibrium after 40 years. West and Marland (2001) suggest that the rate of uptake is constant at 337 kg C ha^{-1} yr^{-1} for the first 20 years and then declines linearly over the next 20 years.

Estimates of carbon uptake by soils in the prairie region of Canada as a result of going from conventional tillage to zero tillage to NT vary from 100 to 500 kg C ha^{-1} yr^{-1} (West and Marland, 2001), assuming an annual carbon flux of −200 to −400 kg C ha^{-1} yr^{-1} in soil organic carbon, plus another −31 kg C ha^{-1} yr^{-1} in saved emissions. The net discounted carbon prevented from entering the atmosphere as a result of a shift to zero tillage from conventional tillage thus depends on the discount rate as indicated in Table 5.5. Total carbon uptake due to agricultural operations varies from about 4 t C per ha to at most 12.5 t C per ha. Compared to plantation forests, the amount of carbon that can potentially be prevented from entering the atmosphere via a dramatic change in agricultural practices is small.

Table 5.5 Expected Annual and Total Carbon Savings from Adopting Zero-Tillage Practices (tonnes of C per ha)

Assumed Annual Sequestration in Soil Organic C during First 20 years	2% Discount Rate		4% Discount Rate	
	Total	Annual	Total	Annual
200	5.94	0.12	4.16	0.17
300	8.13	0.16	5.86	0.23
400	10.33	0.21	7.55	0.30
500	12.52	0.25	9.24	0.37

Source: Manley et al., 2004.

Neglected in the foregoing analysis are carbon leakages. Zero tillage generally leads to lower yields, causing prices to rise. This will cause an expansion in output as grass or forestland is converted to crops and more inputs are employed, and this will lead to greater release of carbon.

Also neglected in this analysis is the potential cost to farmers. With the exception of conversion to grassland, continuous cropping and zero tillage reduce yields and increase production costs because more chemical inputs are required (Lerohl and van Kooten, 1995). For example Lerohl and van Kooten (1995) consider the costs of carbon uptake under continuous cropping of wheat versus a two-year, wheat tillage-fallow rotation. Under the best conditions or most productive land, average annual wheat yields are 2.95 t ha^{-1} under continuous cropping compared with 1.68 t ha^{-1} with the two-year, wheat-fallow rotation, while respective annual net returns are $69.86 and $117.45 per ha. Assuming that organic matter is 50% carbon and that cropping leaves 15%–20% crop residue, the cost of carbon uptake amounts to $373–$498 per t C! This ignores lower levels of C flux with the wheat-fallow rotation since there are fewer machinery operations and less inputs are employed. If the cost difference between conventional tillage and zero tillage is just as great as that between continuous cropping and tillage fallow (about $47.50 ha^{-1}), then, from data in Table 5.5, the cost of carbon uptake via changes in agricultural practices amounts to approximately \$130–\$400 per t C. This is clearly an expensive means for sequestering carbon, and society must have more cost-effective options.

Wood-Product Sinks and Biomass Burning

By producing energy from wood biomass rather than fossil fuels, countries are able to reduce their overall CO_2 emissions by an amount approximately equal to the savings in emissions from the fossil fuels that are replaced. This is because any CO_2 released by burning wood biomass is claimed as a credit by growing the trees that are subsequently burned. Preliminary analysis by van Kooten et al., (2000) suggests, however, that, for a realistic cost of carbon uptake of less than \$20 per t C, the wood-burning option is not likely to be viable, and one would expect very little (marginal) agricultural land to be planted to trees for this purpose. However, these authors show that, if wood is harvested and account is taken of carbon that enters wood-product pools, afforestation of marginal agricultural land could be a useful component of Canada's policy arsenal. They conclude that, for carbon uptake costs of \$20 per t C or less, it may be worthwhile to plant hybrid poplar (but not native species) on maybe one-third of the agricultural land that a non-economist might identify as suitable for afforestation.

If forest plantations and wood-product carbon sinks are a strategy followed by other countries, then we encounter the problem of leakage. For this reason biomass-burning may be a better option for countries with vast forest and agricultural areas. Indeed, the benefits of carbon uptake through afforestation are enhanced under current rules when there exist opportunities to use forest biomass in conjunction with wood waste to produce energy that substitutes for energy from fossil fuels (with the reduction in GHG emissions from fossil-fuel consumption constituting a credit). In terms of carbon balance, the benefits of burning biomass to produce energy include the maintenance of an emissions-uptake equilibrium (no net flux), the one-time gain in carbon uptake from initial establishment of a tree plantation, and the annual offset of fossil-fuel offsetting emissions.

For some countries, biomass burning may be important if the nuclear, wind, and hydropower options are thought to be environmentally unsound and solar power too expensive. In Canada, the forest sector is a large consumer of electricity, much of it purchased from local or regional providers. The purchased electricity is generated from a variety of sources, primarily natural gas, coal, and hydropower. The forest sector self-generates about half of the power that it uses (Forest Sector Table, 1999), but is constrained in many cases from achieving economies of size in power generation by either an inability to sell excess power into the provincial grid or a lack of fibre (Canadian Pulp and Paper Association, 2000). In British Columbia, for example, BC Hydro at one time restricted the sale of privately generated power into the provincial grid because this would reduce prices and the revenues that the government-owned company could generate. Until recently, sawmills in the province burned sawdust in beehive burners, but when this was no longer permitted on environmental grounds, the sawdust was simply put into landfills. A small number of cogeneration plants have been built since the ban, primarily in areas where disposal costs and wood waste volumes are highest.

Firms do not currently pay for wood waste, but its availability may prevent economies of scale in biomass burning. If carbon taxes, subsidies, or government regulations cause firms to take biomass burning more seriously, demand for industrial wood waste will increase beyond current supply, and wood waste will take on value. This, in turn, will encourage production of wood fibre from fast-growing energy plantations. If biomass power generation is determined to be economically profitable, farmers may be able to sell biomass at a profit; at least, it might reduce the compensation (private or public) paid to farmers for establishing and maintaining tree plantations, and increase the area economically feasible for afforestation.

Besides the reduction in CO_2 emissions, biomass burning can provide opportunities for industry and communities to reduce electricity costs if power generators are scaled to their particular requirements. The establishment and operation of biomass systems will increase employment, particularly in rural areas where jobs are most threatened by ongoing forest protection measures and mechanization of factors of production.

Wood-fuel conversion technologies include direct combustion, cogeneration, gasification, and conversion to liquid fuels. The efficiency of the conversion system determines the reduction in CO_2 emissions through the displacement of fossil fuels. Estimates of emission savings range from 1.7 to 9.0 t C ha^{-1} per year depending on forest type, discount rates, energy conversion efficiency, and the particular fossil fuel being displaced (Wright et al., 1992; van Kooten et al., 1999, 2000). The cost of substituting wood biomass for coal in electricity production ranges from $27.60 to $48.80 per t C, based on a value of $7.50 per m^3 for hybrid poplar on energy plantations, a substitution ratio of 2.6–4.6 m^3 of wood per t of coal to generate an equivalent amount of energy, and a carbon content of 0.707 t C per t of coal (Marland et al., 1995).

As shown in Table 5.6, energy from wood residues can compete with fossil fuels and purchased electricity. This conclusion needs careful scrutiny, however. First, wood residue prices are based on average and not marginal costs, and are only available for small-scale operations where wood is easy to come by. At a larger scale, one would expect much higher raw material (wood) costs. Second, wood fibre prices vary significantly by region depending on environmental regulations and residue surpluses or shortages. Regional values are not currently available for comparison (Forest Sector Table 1999). If fast-growing plantations are included, estimated costs are $2.82 per gigajoule (GJ), which is more expensive than fossil fuels but still cheaper than purchased electricity.

Table 5.6 Energy Price Comparisons for British Columbia, Canada ($1997)

Wood Residues	
(assumed conversion factor = 18 GJ per dry tonne)	
Wood residue from pulp and paper mills	$ 1.0 GJ^{-1}
Wood residue from wood industry	$ 0.56 GJ^{-1}
Wood waste plus plantation wood	$ 2.82 GJ^{-1}
Fossil Fuels	
(based on natural gas, boiler efficiency of 85%, C emission factor of 0.050 t GJ^{-1})	
Fuel price	$ 1.73 GJ^{-1}
Electricity	
(at $0.039 per kWh)	$10.84 GJ^{-1}

Source: Forest Sector Table, 1999, and own calculations.

Fossil-fuel substitution on a global scale, using 10% of an estimated 3,454 million ha of forested area as a source for biomass energy, would replace an average of 2.45 Gt C per year. This figure is based on 7 t C ha^{-1} yr^{-1}, while the average carbon capture rate can vary from less than 0.5 to 12 t C ha^{-1} yr^{-1} depending on the type of forestry being practiced—conventional or plantation. This amounts to some 40% of global fossil-fuel emissions of carbon in 1990. In Canada, the high capital cost of infrastructure, regulation of the electricity market, and the relatively low cost of fossil fuels restrict the economic viability of substituting biomass for fossil fuels in power generation. When we consider global climate change, future energy requirements, availability of supply, and social and environmental values, we find that the benefits of renewable energy sources such as wood biomass outweigh the costs in some, but not all situations.

Conclusions

It is estimated that terrestrial vegetation and soils currently absorb the equivalent of 40% of global CO_2 emissions that come from human activity. This could be increased significantly by changes in land use, particularly by increased planting of forests and reduction of deforestation. However, the ability to increase the uptake of carbon in terrestrial ecosystems through these options is limited, with any carbon stored in this manner likely to be ephemeral and subject to the vagaries of socio-economic and political variables. Further, more intensive management of land, including the increased application of fertilizers, could result in increased emissions of other greenhouse gases, such as methane and nitrous oxide, that have a much greater warming potential than CO_2. This would negate any benefits accrued from increased CO_2 fixation. Perhaps the best terrestrial option is greater use of biofuels that substitute for fossil fuels in energy production. Again, the economic attractiveness of this option is likely to be limited.

Only technological innovation and the restructuring of energy generation and consumption (as outlined in chap. 6) can provide solutions to the problem of reducing greenhouse gases in the atmosphere. Land-use changes may have the potential to provide a short-term bridge to enable these solutions to be implemented, but that too may be optimistic.

References

Adams, D.M., R.J. Alig, B.A. McCarl, J.M. Callaway, and S.M. Winnett. (1999). Minimum Cost Strategies for Sequestering Carbon in Forests. *Land Economics* 75: 360–74.

Betts, R.A. (2000). Offset of the Potential Carbon Sink from Boreal Forestation by Decreases in Surface Albedo. *Nature* 408 (9 November): 187–90.

Bromley, D.W. (1999). *Sustaining Development: Environmental Resources in Developing Countries.* Cheltenham, UK: Edward Elgar.

Callan, B.E. (1998). *Diseases of "Populus" in British Columbia: A Diagnostic Manual.* Victoria, BC: Natural Resources Canada, Canadian Forest Service.

Canadian Pulp and Paper Association. (2000). CPPA's Discussion Paper on Climate Change. Montreal, QC: CPPA. <www.open.doors.cppa.ca>

Forest Sector Table. (1999). *Options Report: Options for the Forest Sector to Contribute to Canada's National Implementation Strategy for the Kyoto Protocol.* Ottawa: Natural Resources Canada, Canadian Forest Service, National Climate Change Program.

IPCC (Intergovernmental Panel on Climate Change). (2000). *Land Use, Land-Use Change, and Forestry.* Cambridge, UK: Cambridge University Press.

IPCC (Intergovernmental Panel on Climate Change). (2001). *Climate Change 2001: Mitigation. Technical Summary.* Contribution of Working Group III to the Third Assessment Report. Cambridge, UK: Cambridge University Press.

Krcmar, E., and G.C. van Kooten. (2001). *Timber, Carbon Uptake and Structural Diversity Tradeoffs in Forest Management.* FEPA working paper. Vancouver: BC.

Krcmar, E., B. Stennes, G.C. van Kooten, and I. Vertinsky. (2001). Carbon Sequestration and Land Management under Uncertainty. *European Journal of Operational Research* 135 (December): 616–29.

Lerohl, M.L., and G.C. van Kooten. (1995). Is Soil Erosion a Problem on the Canadian Prairies? *Prairie Forum* 20 (Spring): 107–21.

Manley, J., G.C. van Kooten, K. Moeltner, and D. Johnson. (2004). Creating Carbon Offsets in Agriculture through Zero Tillage: A Meta-Analysis of Costs and Carbon Benefits. *Climate Change* (in press).

Marland, G., T. Boden, and R.J. Andres. (1995). Carbon Dioxide Emissions from Fossil Fuel Burning: Emissions Coefficients and the Global Contribution of Eastern European Countries. *Quarterly Journal of the Hungarian Meteorological Service* 99 (July–December): 157–70.

Moffat, A.S. (1997). Ecology: Resurgant Forests Can Be Greenhouse Gas Sponges. *Science* 277: 315–16.

Papadopol, C.S. (2000). Impacts of Climate Warming on Forests in Ontario: Options for Adaptation and Mitigation. *The Forestry Chronicle* 76 (January/February): 139–49.

Paustian, K., O. Andren, H.H. Janzen, R. Lal, P. Smith, G. Tian, H. Thiessen, M. van Noordwijk, and P.L. Woomer. (1997). Agricultural Soils as a Sink to Mitigate CO_2 Emissions. *Soil Use and Management* 13: 203–44.

Read, D., D. Beerling, M. Cannell, P. Cox, P. Curran, J. Grace, P. Ineson, P.G. Jarvis, Y. Malhi, D. Powlson, J. Shepherd, and I. Woodward. (2001): The Role of Land Carbon Sinks in Mitigating Global Climate Change. London: Royal Society, 6.

Richey, J.E., J.M. Melack, A.K. Aufdenkampe, V.M. Ballester, and L.L. Hess. (2002). Outgassing from Amazonian Rivers and Wetlands as a Large Tropical Source of Atmospheric CO_2. *Nature* 416: 617–20.

Schimel, J.I., K.A. House, P. Hibbard, P. Ciais Bousquet, P. Peylin, B.H. Braswell, et al., (2001). Recent Patterns and Mechanisms of Carbon Exchange by Terrestrial Ecosystems. *Nature* 414: 169–72.

Sedjo, R.A., J. Wisniewski, A.V. Sample and J.D. Kinsman. (1995). The Economics of Managing Carbon via Forestry: Assessment of Existing Studies. *Environmental and Resource Economic* 6: 139–65.

Sohngen, B., and R. Sedjo. (1999). *Estimating Potential Leakage from Regional Forest Carbon Sequestration Programs.* RFF Working Paper. Washington, DC: Resources for the Future.

Stennes, B. (2000). *Carbon Uptake Strategies in the Western Boreal Forest Region of Canada: Economic Considerations.* PhD Diss., Department of Forest Resources Management, University of British Columbia.

van Kooten, G.C. (2000). Economic Dynamics of Tree Planting for Carbon Uptake on Marginal Agricultural Lands. *Canadian Journal of Agricultural Economics* 48 (March), 51–65.

van Kooten, G.C., E. Krcmar-Nozic, B. Stennes and R. van Gorkom. (1999). Economics of Fossil Fuel Substitution and Wood Product Sinks when Trees are Planted to Sequester Carbon on Agricultural Lands in Western Canada. *Canadian Journal of Forest Research* 29(11), 1669–78.

van Kooten, G.C., B. Stennes, E. Kremar-Nozic, and R. van Gorkom. (2000). Economics of Afforestation for Carbon Sequestration in Western Canada. *The Forestry Chronicle* 76 (January/February): 165-72.

van Kooten, G.C., S. Shaikh, and P. Suchánek. (2002). Mitigating Climate Change by Planting Trees: The Transaction Costs Trap. *Land Economics* 78 (November): 559-72.

van Kooten, G.C., W.A. Thompson, and I. Vertinsky. (1993). Economics of Reforestation in British Columbia when Benefits of CO_2 Reduction are Taken into Account. In *Forestry and the Environment: Economic Perspectives.* W.L. Adamowicz, W. White and W.E. Phillips, eds. 227-47. Wallingford, UK: CABI Publishing.

Watson, R.T., M.C. Zinyowera, R.H. Moss, and D.J. Dokken. (1996). *Climate Change 1995: Impacts, Adaptations and Mitigation of Climate Change: Scientific-Technical Analysis.* Cambridge, UK: Cambridge University Press.

West, T.O., and G. Marland. (2001). A Synthesis of Carbon Sequestration, Carbon Emissions, and Net Carbon Flux in Agriculture: Comparing Tillage Practices in the United States. Environmental Sciences Division Working Paper. Oak Ridge, TN: Oak Ridge National Library.

Wilson, B., G.C. van Kooten, I. Vertinsky, and L.M. Arthur, eds. (1999). *Forest Policy: International Case Studies.* Wallingford, UK: CABI Publishing.

Woomer, P.L., C.A. Palm, J. Alegre, C. Castilla, D. Cordeiro, K. Hairiah, J. Kotto-Same, et al., (1999). *Carbon Dynamics in Slash-and-Burn Systems and Land Use Alternatives: Findings of the Alternative to Slash-and-Burn Programme.* Working Paper. Nairobi, Kenya: Tropical Soil Biology and Fertility Program.

Wright, L.L., R.L. Graham, A.F. Turhollow, and B.C. English. (1992). The Potential Impacts of Short-Rotation Woody Crops on Carbon Conservation. In *Forests and Global Climate Change,* vol. 1, ed. D. Hair and R.N. Sampson. Washington, DC: American Forests.

Technology and Climate Change

Gerard F. McLean and Murray Love

Introduction

According to the IPCC, there is now persuasive evidence that anthropogenic emissions of carbon dioxide are contributing to global climate change. The major fraction of these emissions is generated by the fossil-fueled energy conversion technologies employed by industrial societies since the onset of the Industrial Revolution. Humans have come to rely on commercial energy technologies to provide an ever-expanding array of goods and services, but it is these very technologies that now threaten to cause irreversible damage to natural systems and human societies. Climate change is therefore a consequence of the advance of a high-technology society—a process that will certainly accelerate in the coming decades—and it prompts us to consider the role of technology, both in creating the problem and in potentially solving it. Our aim in this chapter is to explain the links between human development, technology, and climate change, and to review some of the issues society must face if we decide to adopt carbon-free technologies while maintaining our standard of living and increasing that of the developing world.

Technology and Energy

Although it is easy to overlook, nearly all technological activity relies on large-scale energy conversion. Energy is used in the design, manufacture, and distribution of a technology and often in its use; every car, for instance, contains the embodied energy of countless smelting, refining, and manufacturing processes, at the same time as it uses the energy stored in gasoline to operate. Even the bland surface of the pencil, that exemplar of "appropriate" low technology, conceals the logging operations, graphite mines, and castor oil refineries required to produce it. Our technological services are unbreakably linked to large-scale energy use, and these in turn rely for the most part on the combustion of hydrocarbon fuels. The question facing us at the beginning of the twenty-first century is whether the link between energy and climate change is as unbreakable as that between technology and energy use. Can we wean ourselves from hydrocarbon energy sources without crippling the world economy—a prospect as deadly as any climate change scenario?

All of society's services—among them heat, light, information, transportation, nutrition and hygiene—are provided by the technologies that comprise these sectors. Figure 6.1 shows a breakdown of global energy use by energy source. In 1999, the world consumed about 402 exajoules (1 EJ = 10^{18} J) of commercial energy, of which over 85% was due to the combustion of petroleum, coal, and natural gas (EIA 2002). Only 58.4 EJ (14.5%) came from carbon-free sources, and of these, renewable sources (comprising geothermal, solar, wind, wood and waste technologies) contributed just 3.0 EJ, 5% of the carbon-free sources and 0.7% of total energy consumption. Nuclear fission and hydroelectric generation made the largest contributions among the carbon-free technologies, generating 26.6 EJ and 28.8 EJ respectively.

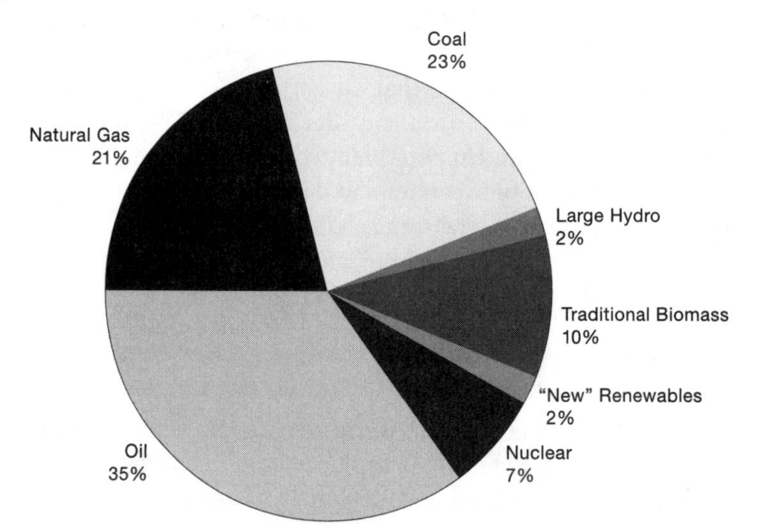

Figure 6.1. Global Commercial Energy Supply by Source
Source: IEA (2000).

Hydrocarbon energy sources have formed the basis for all industrialization to date, due to their impressive technical qualities: high energy density, safe handling, and ease of extraction. Today's developing countries not unreasonably expect to use them as a basis for improvements in their living standards, as all advanced societies have done to date. Although there are numerous technologies for the production of services from hydrocarbons, the majority of these use combustion to drive heat engines, with the resultant emission of waste gases and heat into the atmosphere. Aside from the environmental damage caused by these emissions, combustion of hydrocarbons also wastes these very valuable chemical feedstocks, from which so many of our products are made. Combustion is the culprit.

Options

Given the problematic linkage between technology, energy, and greenhouse gas production, there are three broad approaches to the problem:

- *Reduce the demand for services:* reduce emissions through voluntary conservation and improved efficiency.
- *Clean up the emissions after the fact:* continue with combustion as the dominant source of energy but develop new technologies to catch the emissions in the tailpipe or smokestack, or to remove carbon from the atmosphere.
- *Develop carbon-free energy technologies:* develop new post-combustion technologies to break the link between energy services and greenhouse gas emissions.

Regardless of the specific technologies and energy sources used, human development appears to be strongly correlated with energy use. Figure 6.2 displays the relationship between per capita energy use and gross domestic product (GDP) for some 200 or so countries. We also group these countries according to the United Nations Human Development Index (HDI), which classifies countries as high, medium, and low development based on such measures as education levels, literacy, infant mortality, and life expectancy (UNDP, 2001). There is a very strong positive correlation between human development (as measured by either GDP or the HDI) and energy use per capita, and it seems that no countries to date have achieved high levels of development without relying on highly energy-intensive activities, usually in the form of fossil-fuel technologies.

Over the next decades, many of today's developing countries—India and China among them—are likely to move into the ranks of the devel-

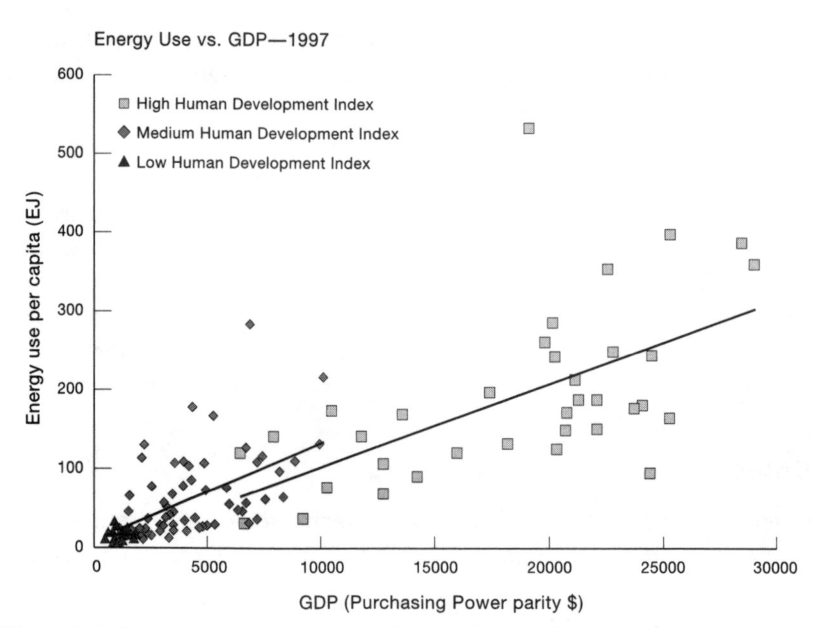

Figure 6.2. Comparison of Energy Use Per Capita and Gross Domestic Product (1997)

Source: UNDP (2001), IEA (2000).

oped, and they will surely demand the technological services we have come to take for granted. Barring unprecedented breakthroughs in the efficiency with which services are supplied, there is simply no question that worldwide energy demand will continue to increase. The International Energy Agency has considered a number of scenarios of future global development, and has come to the conclusion that global energy consumption will increase at a rate of about 2% per year for the foreseeable future. By 2020, the world will use over 640 EJ of commercial energy, a 60% increase over today's levels. In the developed world, energy demand is expected to grow at a slightly slower rate of around 1.7%; the developing countries, containing the majority of the world's population, will grow at about 2.7% (IEA, 2000). It is important to note that *all* the scenarios considered in this study featured a considerable increase in worldwide energy use over today's levels, including those based on strong conservation and efficiency measures.

Energy efficiency and conservation measures are often mentioned as having huge potential to reduce greenhouse gas emissions, but in almost all cases the actual potentials are far lower than commonly supposed. In 1999, Canada developed the National Options Tables Exercise in order to assess the potential efficiency and conservation mechanisms to achieve reductions in greenhouse gas emissions in a variety of economic sectors (Morrison, et al., 2002). In general, two conclusions remained valid for all sectors. The first was that there was significant potential for short-term reductions in greenhouse gas emissions through measures to increase efficiency and reduce energy use. The second was that these reductions would be nowhere near large enough to achieve any reduction targets and that further reductions would have negative economic impact. Both global and national studies for future energy scenarios reach the same conclusion: measures to curb energy use through reduced demand for services or to improve the efficiency with which services are delivered will not be sufficient to stabilize emissions, and may even lead to increased energy use (Herring, 1999).

Carbon-sequestration technologies seem to offer an easy way to avoid the threat of climate change, allowing us to continue to expand our use of hydrocarbon fuels in a "sustainable" manner. Unfortunately, these technologies do not yet exist in an economically viable form, although there are many projects in development. "Zero-emission coal" uses gasification and scrubbing to extract hydrogen from coal, sequestering the carbon products into mineral deposits (USDOE, 1999). Many other approaches to carbon sequestration are not as promising, often using up to 30% of the energy produced by the hydrocarbon in order to remove the carbon. This translates approximately into a 50% increase in the cost of energy-related services, an economically daunting prospect.

Of the three options for dealing with carbon emissions, demand reduction and emission clean-up are not expected to provide any significant impact in the face of continued and expanded demand for energy services. The third option—the development of post-combustion technology that simply does not emit greenhouse gases—warrants further consideration.

Sources for Zero-Carbon Energy

No matter what technologies are used to manipulate energy, the link between technological services and environmental degradation is dependent on the initial source of energy. As we have shown, the major fraction of our current energy services is derived from conventional oil, followed by coal and natural gas. Our suite of technically feasible carbon-free energy technologies is limited to the renewables, a category including geothermal, solar, wind, wood and waste sources, hydroelectric power, and nuclear fission.

It is difficult at first to comprehend the relative magnitudes of energy delivered by these different sources. An interesting comparison between power-generation technologies is found in the Three Gorges Dam project in China. This is the world's largest hydroelectric installation, a 600-km-long reservoir in the Yangtze River inundating 632 km^2 of land, forcing the resettlement of over one million people and the flooding of 1,300 heritage sites. This massive project will generate 18.2 gigawatts (GW) of electricity, an amount equivalent to the electricity produced by 18 large nuclear plants, the burning of 400 million tonnes of coal annually, or the annual electrical production of about 54 million large wind turbines.

Renewable sources of energy are beginning to make inroads, especially in Europe and North America, while hydroelectricity and nuclear power are both significant and proven players in our current energy system. However, both nuclear fission and hydroelectric generation are widely opposed in the industrialized world due to negative perceptions of their environmental and safety characteristics, and to the runaway expenses of the first generation of nuclear reactors. The renewables are generally perceived to be environmentally benign and sustainable.

Renewable energy sources

Renewable energy—primarily wind, solar, tidal, and wave generation systems—generate electricity by extracting power from naturally occurring energy flows. There are no emissions associated with renewables, and even those based on combustion of biomass produce no net emissions by virtue of closing the carbon cycle through the atmosphere. The capital costs of renewables are steadily declining, and many jurisdictions worldwide are beginning to legislate the inclusion of renewable sources into their energy mix.

Renewable energy sources (excluding hydroelectric generation) currently account for only 2% of global energy production. Biomass technologies (largely wood and waste combustion) make up the major component of renewable generation. However, most of these are attached to industrial installations such as sawmills making use of wood waste, and there is little prospect for large-scale growth (EIA, 2002). While renewable technologies are experiencing remarkable growth and popularity, it is not clear that their widespread adoption will necessarily follow. Renewable sources pose serious technical challenges. Also, as the penetration of renewables increases, it is reasonable to expect public concern over their impact will also increase. To anticipate these issues, we begin to consider some of the unique difficulties posed by renewable energy technologies.

Renewable energy sources are intermittent and largely random; while we are able to predict how much energy we should be able to derive from a renewable source over a lengthy period of time, we cannot predict precisely *when* that energy will be produced. Renewables therefore always require backup energy storage or generation of equal capacity to guarantee that society's services will continue to be delivered. To put it another way, every kilowatt of renewable capacity requires a kilowatt of conventional generation (or large-scale storage) as backup. Renewables based on solar and wind energy operate only 15–35% of the time, meaning that the conventional "backup" would actually be the primary energy generator. Without backup or storage technologies in place, an energy system based on renewables would experience lengthy and frequent blackouts.

One approach to dealing with this intermittency is to rely on the electricity grid to provide support for the renewable facilities. Denmark boasts that wind energy supplies about 14% of its electricity needs. However, there is growing evidence that the Danes have made substantial economic and technical sacrifices in order to meet this goal. One report claims not only that wind power has been a large net economic burden on the country, but that up to 40% of Denmark's wind electricity is generated at periods of low demand and is therefore "dumped" (sold below cost) on the Northern European market (LNtV, 2002). Other analysts estimate that the grid can accept up to 20%–25% of unpredictably intermittent generation before the system is destabilized, causing blackouts (Laughton and Spare, 2001). Thus, while renewable sources today can reduce conventional fuel consumption, they cannot easily displace conventional installed capacity. To assume that an isolated but grid-connected renewable generation facility is a prototype for a future clean energy system is an overly optimistic leap of faith.

Renewables must harvest extremely dilute energy flows, and as a result are at least an order of magnitude more land-intensive than any other energy technology, with the possible exception of large hydroelectric installations (Smil, 1984). In most cases, renewables are even more land-intensive than the human societies they supply, meaning that the energy technology alone requires more land than the cities and towns it is powering. Some attempts have been made to tackle this question, for instance, by claiming that the entire 1997 US electricity demand could be satisfied by a square area of photovoltaic panels 100 miles on a side situated in Nevada (Turner, 1999). Unfortunately, these analyses are rather too simplistic, implicitly assuming (among other things) that a perfectly efficient storage and transmission system of unlimited capacity will be available to carry the system through indefinitely long periods without sun or when demand exceeds supply. The actual area required is likely to be many times larger, and this question is the focus of ongoing research within IESVic (Niet, 2001).

As renewable installations become common, they will face increasing competition for land, both from alternative non-energy developments and from preservationist campaigns. Wind developers in the UK are already facing stiff local opposition to their plans to build large wind farms in scenic areas of Wales, at a time when wind supplies less than 1% of the UK's electricity. In Denmark, wind developers have almost run out of feasible land-based sites, and are now looking to offshore installations, which tend to be 50–100% more expensive to build and maintain (EIA, 2002). Wind farms in Alberta's Pincher Creek account for only a tiny fraction of the province's electricity supply; it would take over 700 such projects (not including storage and backup) to even approximate the size of the province's conventional generating capacity—this in the Canadian province most suited for large-scale wind development.

While the potential for renewable energy is huge, it may turn out that the public will be far more resistant to renewable energy projects once they grasp the sheer magnitude of the installations required to generate any truly significant amount of electricity. This issue will be compounded in a future society in which renewables would be required to provide a greater amount of energy than all today's power plants and cars combined. Renewable technologies can and must make a contribution to our future energy needs, but we need to realize that they are just one set of energy technologies out of many. Renewable energy will grow in importance as we move towards zero-carbon energy sources, but it is doubtful they will satisfy the entire demand for new energy. Renewables alone will not be enough.

Hydroelectric generation

Hydroelectricity is still included in the "renewable" category for many energy forecasts, but in practice has fallen out of favour with environmentalists and advocates of renewables. Hydroelectric generation is very clean, and—where feasible—is capable of supplying large amounts of electricity, but dams destroy watersheds and interfere with fisheries. Furthermore, many of the best hydroelectric sites in the industrialized world have already been developed. As a result, there is little potential for future hydroelectric expansion in the industrialized world; Canada alone is planning new hydroelectric capacity. The developing world, however, continues to build new hydroelectric facilities; predictions suggest these regions will add about 50% to the existing worldwide hydroelectric capacity by 2020 (EIA, 2002). However, large-scale hydroelectricity does appear to be experiencing declining returns, and it is unlikely that the world will see sustained growth in these facilities.

Worldwide, hydroelectric generation accounted for about 7% of energy consumed in 1999, comparable in magnitude to nuclear power, but equivalent to only one-third of the energy provided by natural gas or coal, and less than one-fifth of the energy supplied by petroleum. Canada has one of the world's highest proportions of hydroelectric facilities, generating 60% of the country's electricity in 1999, and 20% of the total energy consumed. Unfortunately, hydroelectricity is extremely site specific. For instance, while the Three Gorges Dam project in China is impressive, it is not easily repeatable; it would be possible to double output from other sources by building another 18 nuclear plants or burning another 400 million tonnes of coal each year, but there is only one Yangzte River, and there are few if any other sites that match its potential. Geography ultimately limits the potential for hydroelectric power.

In regions where hydroelectric power is plentiful, it is a powerful and flexible energy technology, capable of providing both base- and peaking-load generation as conditions require. Hydroelectricity may be an ideal buffer technology for intermittent renewables, since it can be taken on- or off-line at very short notice, thus providing the storage capacity needed to allow the intermittent renewable sources to satisfy a regular, periodic demand cycle.

Nuclear energy

Nuclear fission was responsible for 12.3% of Canada's electricity use in 1999, and for about 6% of total energy consumption. Worldwide, nuclear power supplied about 6% of total commercial energy demand. Nuclear fission does not involve combustion of any kind and therefore causes no airborne emissions. All of the spent fuel from a nuclear reactor is contained; none is released to the environment, and the volume of waste is equal to the volume of the fuel. Since nuclear reactors require very small amounts of fuel (on the order of a few truckloads every couple of years) the volume of waste is likewise small. From the viewpoint of the electric utilities and their customers, nuclear energy is almost indistinguishable from coal—both are baseload technologies, producing vast quantities of electricity around the clock—and indeed, nuclear energy in Canada displaces coal generation, avoiding about one million tonnes of CO_2 emissions every week (Whitlock, 2002). These factors have led Canada to argue for its installed nuclear generating capacity to be counted for carbon credits in its negotiations around the Kyoto Protocol. Canada's official stance is pro-nuclear.

However, nuclear energy is also the most controversial energy source: its first generation has been largely uneconomical, the risks of radiation release from accidents or waste disposal makes it too risky in the public eye, and the potential to facilitate nuclear weapons proliferation makes it problematic in a post 9/11 era. The technical validity of these concerns is hotly contested between an industry that is seen to have a strong vested interest, and a concerned public that is suspicious of "big energy."

Certainly the first generation of nuclear reactors too often experienced severe cost overruns, due to a combination of industry mismanagement and poorly addressed public concerns (Rhodes, 1993). However, the nuclear industry now claims that a new generation of nuclear reactors will compete head-on with natural gas generation. Meanwhile, the first generation of nuclear reactors continues to improve in performance, often producing astonishingly cheap electricity, and many are receiving licence extensions of 20 years or more. Further, the volume of high-level nuclear waste is very small, and is containable, unlike the vast and uncontained waste dumped into the atmosphere through the burning of fossil fuels. In Canada, waste disposal and reactor decommissioning costs are included in the price of nuclear electricity. The major externalities of nuclear power are therefore fully internalized—making it the only large-scale generation technology to do so.

Since the first civilian power reactors were commissioned in the 1950s, not one person has been killed by radiation releases from western civilian nuclear power plants. Even the catastrophe at Chernobyl—surely the worst-case accident for nuclear power—had a far smaller death toll than is commonly assumed, killing 35 within a few days and fewer than two dozen more in the 16 years since (UNSCEAR, 2000).

While the controversy surrounding nuclear energy will not soon abate, it does have an enormous potential to provide levels of baseload energy suitable to an advanced industrial society, and to do so using a very small amount of land. France has reduced its carbon emissions drastically since its adoption of nuclear power, which now provides over 75% of its electricity generation (Nakicenovic, 1996). However, the nuclear industry has not yet shown that it can build reactors competitive in price with natural gas generation, and with the inherent safety features that would be sufficient to address public concerns over the technology. Finally, nuclear energy is still a baseload technology, running most efficiently at a constant level of output; current designs are not able to follow loads to meet peaking demands. This means that any carbon-free energy system based on nuclear energy would require some form of complementing storage or peak load generation technology to be installed in tandem with the nuclear source.

Post-Combustion Technology

This review of carbon-free energy sources does not paint an optimistic picture of the future of the energy system and the societies that depend on it. The hydrocarbon sources on which our current system depends are now recognized as the primary driver of global climate change. Nuclear power is regarded by many as simply too hazardous; furthermore, current reactor designs can generate only baseload electricity, an incomplete contribution to the overall energy mix. Energy from "new renewables" is dispersed and intermittent, and is therefore incapable of providing the round-the-clock service availability we require. Modern biomass will require extensive use of arable land for so-called energy crops, displacing land currently used for food crops or recreation. We are swimming in a sea of energy, but there is no single clear alternative to the status quo that will provide the services we demand without significant negative impacts. However, in spite of these difficulties there are pathways to clean energy that can be developed through technological innovation.

The energy system involves the delivery of energy-intensive services using energy derived from a source and stored in an energy carrier, or "currency." A schematic diagram of the chain of technological connections linking services back to sources is shown in Figure 6.3. The currencies we use to carry energy value are usually tailored to the particular needs of the service. For instance, automobiles operate on gasoline (an energy currency made from oil) whereas household lighting operates on electricity, which can be made from almost any arbitrary energy source: oil, coal, natural gas, uranium, biomass, wind, or sunlight, to name a few. Oil and electricity are therefore fundamentally different energy currencies with fundamentally different characteristics. Together, the dual currencies of electricity from arbitrary sources and hydrocarbon fuels from oil and gas provide the mix for the current energy system. This system, while utilizing significant non-hydrocarbon source technologies, is still based on hydrocarbon (and hence combustion). This presents a huge problem for the prospect of weaning the world off hydrocarbons: they are tremendously effective at storing large quantities of energy.

A vast array of energy services relies wholly on stored energy. In road transportation, all propulsion energy must be stored on board the vehicle, creating a demand for highly efficient energy currencies containing high amounts of stored energy for a given volume and weight of fuel. In stationary applications, the need for energy storage is less evident but nonetheless critical. When a utility experiences a sudden increase in electricity demand, it must have access to an instantly available electricity supply. In

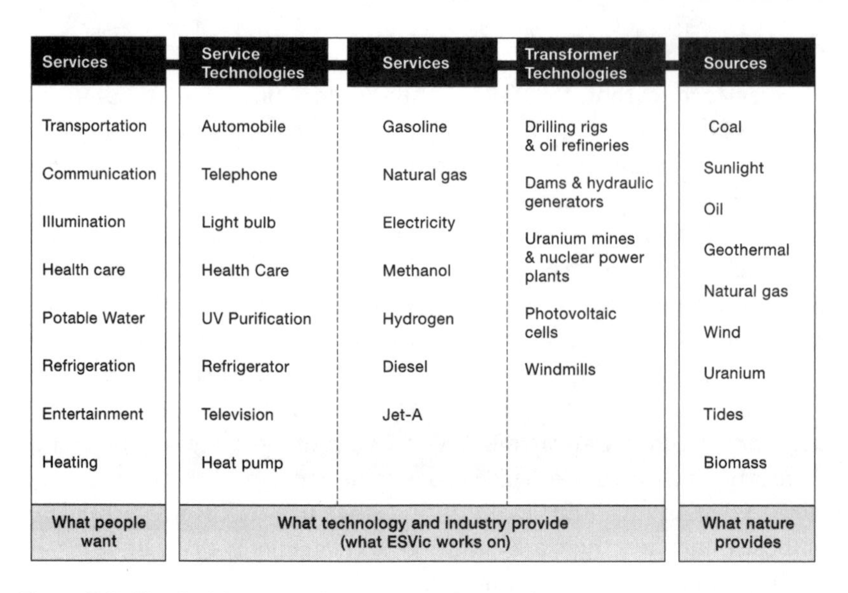

Figure 6.3. The Architecture of the Energy System

Source: Service Energy Linkages. Institute for Integrated Energy Systems, University of Victoria (IESVic).

practice, this is achieved by using stored hydrocarbon currencies (such as diesel or natural gas) in generators with fast time response to augment slow-response baseload capacity. It is much more difficult to recover the excess energy that is dumped when demand suddenly decreases. This would require the ability to re-store the excess energy, and unfortunately hydrocarbons provide a one-way flow of energy only. They are not reversible: gasoline can be converted into electricity, but electricity cannot be converted into gasoline. Unfortunately, none of the candidates for carbon-free energy production discussed in the previous section can provide an easy path to currencies that store energy. They all generate electricity (or heat) only. To create a long-term sustainable energy system that can use the energy offered by carbon-free energy technologies, we require an energy currency that allows energy to be stored, and that can be created from both electricity and heat. If it were possible to create an energy currency from electricity that could be stored, transported, and distributed in different forms, we would begin to have a truly viable alternative to hydrocarbon currencies. Such a currency would begin to usher in a post-combustion age.

Fortunately, there is such a currency: hydrogen. It has the highest energy density by weight of any fuel. Hydrogen can be made from any energy source, using thermal, chemical, biological, or electrolytic processes.

It can do work through direct combustion or (much more efficiently) through electrochemical conversion to electricity in a fuel cell. There are large technical and economic barriers to a hydrogen-based energy system, but hydrogen is the only alternative for the future that will allow us to continue developing both stationary and transportation energy services without increasing the risk of large-scale environmental damage. Hydrogen is source-independent and, together with carbon-free electricity, it can provide all the services we enjoy today in an environmentally benign manner. While it is unlikely that we will ever drive wind-powered cars, it is most likely that we will drive hydrogen-fuelled cars, with the hydrogen produced from wind-generated electricity.

Since hydrogen can be produced from any energy source, it makes sense to integrate energy production for both stationary and transportation services to produce the dual currencies of electricity and hydrogen synergistically. Figure 6.4 provides a schematic block diagram of this emerging system. Using this approach, many of the difficulties identified with carbon-free energy technologies can (at least in principle) be addressed.

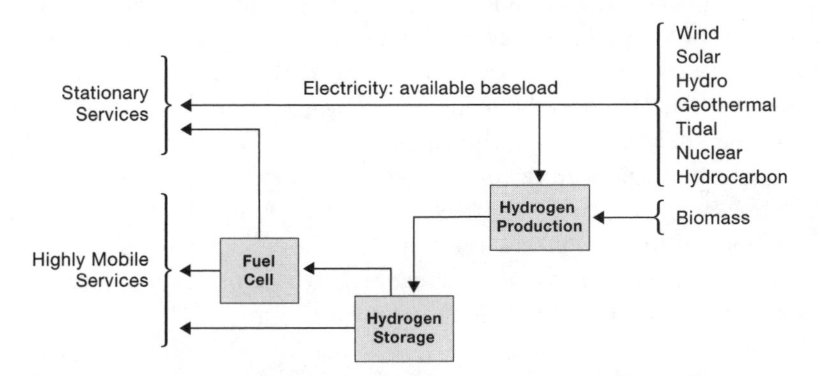

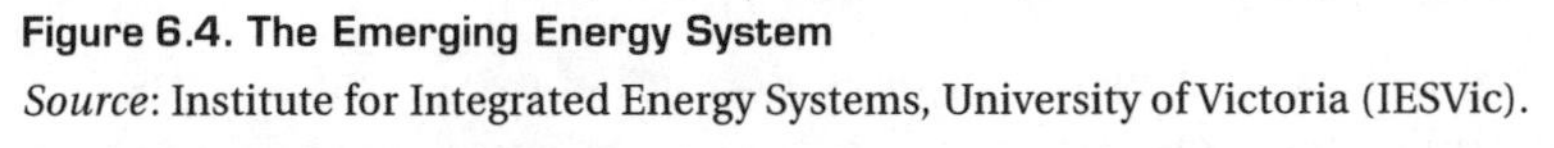

Figure 6.4. The Emerging Energy System

Source: Institute for Integrated Energy Systems, University of Victoria (IESVic).

Current nuclear power plants, as mentioned, can only output constant levels of electricity. In this case, hydrogen production can be used as a buffering element: the nuclear plant would be designed to produce an oversupply of electricity, and the excess would be used for hydrogen production and storage. The process would be controlled to provide the dynamic characteristics required for load following electrical service supply. Instead of trying to store the produced electricity, the plant would produce a mix of both electricity and hydrogen, with the rate of hydrogen production being controlled to "throttle" the electrical output.

The combination of hydrogen production and electricity generation from hydrogen will allow renewable sources to deliver guaranteed power for stationary applications. However, such a renewable electricity generation system may be prohibitively expensive, due to the inherent energy cost of transforming one energy currency to another. It seems more likely that renewable sources will be used mainly to produce high value transportation fuel, where there is always a temporal lag between fuel production and use. Instead of competing with installed electrical power generation, the renewables will be competing with alternative forms of transportation fuel production. Even hydrocarbon sources such as coal can benefit from the production of hydrogen since this would provide a centralized, large-scale opportunity to sequester emissions while still allowing the delivery of both stationary and transportation energy services. In fact, a likely path toward hydrogen will begin with the use of conventional hydrocarbon sources to produce hydrogen, followed by sequestration technologies, followed by an eventual transition to carbon-free energy technologies.

In short, it is only through the development of hydrogen technologies that we can achieve independence between energy sources and energy services, and only by breaking the link between energy sources and services that we can begin to wean the planet from hydrocarbons. Any efforts to reduce or make more efficient our use of hydrocarbons are ultimately no more than delaying tactics. Our mission as a society is to create viable energy technologies centred on hydrogen, and then to move towards carbon-free energy sources for combined electricity-and-hydrogen-generation. Technology development is at the core of this mission, but it must be supported by the will to move in this direction as well as by the honest admission that climate change is a technological problem.

Hydrogen technology

The vision of a hydrogen economy outlined above provides a historic opportunity to decouple energy use from greenhouse gas emissions. It is only through the development of hydrogen systems that continued economic growth can be made sustainable. Although there is a vast array of technologies required to implement large-scale hydrogen systems, there are three building blocks on which any hydrogen system must be built:

- There must be some way of generating the hydrogen.
- Once generated, there must be some way of storing the hydrogen.
- To provide services, there must be some way of extracting the energy stored in the hydrogen.

Before proceeding to an overview of the technological progress in each area, it is worth pointing out that Canadian technology is at the forefront of world development in hydrogen systems. Just like electricity, hydrogen does not exist in a rough form as oil does—it must be made. The fastest pathway to hydrogen production will be hydrocarbon reforming processes (Petterson and Westerholm, 2001).

Electricity can be used to generate hydrogen from water through the process of electrolysis. Industrial-scale electrolysis has been producing extremely high purity hydrogen for almost 100 years, so this technology is not new. What is new is the drive to produce hydrogen at a sufficiently low cost to make it cost-competitive with other fuels in an unregulated competitive market. Depending upon the size of the electrolyzer and the method of manufacture, electrolysis provides a near-term means of producing hydrogen with overall efficiencies in excess of 80%. High-grade heat from nuclear reactors can also be used to produce hydrogen (thermolysis). In the case of nuclear power, this would commit the generator (or a portion thereof) to hydrogen production. As mentioned above, a hydrogen-electric nuclear facility could potentially produce large amounts of hydrogen and also provide the nuclear plant with some load-following capabilities. However, there is no reason to discount the possibility of integrating both thermal and electrolytic production within a nuclear generating facility.

It is also possible to produce hydrogen directly using genetically engineered microbes to break down organic waste material. Biomass processing is also a promising alternative to reformation for naturally occurring hydrocarbons such as methane or fuel crops, but is still in the very early stages of research (Mann et al., 1998).

In hydrogen production, all forms require the hydrogen atoms to be gathered from some naturally occurring material and formed into pure elemental hydrogen. No matter what source of energy is used, all known processes for generating hydrogen require large quantities of high purity water. Although proponents of hydrogen tend to ignore this, we point out that high quality water is in scarce supply in many parts in the world, and must compete with agriculture and other uses.

The technologies for hydrogen storage include liquefaction, compression, or adsorption onto a substrate. All three are under aggressive development, and all three are suited to particular applications for hydrogen storage. Although hydrogen has the highest energy density by weight of any fuel, its energy density by volume is quite low, and hydrogen storage tanks tend to be large. Hydrogen therefore compares unfavourably with hydrocarbon fuels such as gasoline on a direct volume comparison. The technology targets currently driving hydrogen storage technology development are those set by the US Department of Energy and the Japanese World Energy Network, both of whom have identified requirements to meet automotive needs. Liquefied hydrogen is not being pursued aggressively in North America. In Europe, particularly in Germany, it is used both for transportation of hydrogen from generator to distribution site and as a direct fuel in internal-combustion applications. The cost of liquefaction is high, since 40% of the energy stored must be used to liquefy the gas at cryogenic temperatures. While some argue the exergy of cold can be used to recover this energy investment, such an approach would eliminate the independence between source and service that is one of hydrogen's greatest attributes. Further problems with liquefaction include the inevitable boil-off of hydrogen gas that must be dealt with in any working system.

Compressed hydrogen is the only mature technology for hydrogen storage, having been able to make use of the great improvements in compressed-gas storage for different applications over the past 20 years. Current technology involves plastic- or metal-lined cylinders wrapped with carbon fibres to produce sufficient strength to withstand the extremely high pressures required to make hydrogen into an acceptably low volume for vehicular applications. Conformal tanks, custom-designed to fit in difficult spaces such as car roofs and side panels, are another means of providing on-board storage requirements, in spite of hydrogen's comparatively low volumetric energy density.

The highest volumetric storage density for hydrogen is achieved by adsorbing hydrogen onto the surface of a metal hydride. However, these systems tend to be very heavy, achieving only 1–2% hydrogen by weight in low-temperature systems. Up to 7% hydrogen by weight can be achieved with high-temperature systems that require the application of substantial amounts of heat to desorb (release) the hydrogen. In addition, both approaches tend to have limited lifetimes, with significant degradation in storage performance as the metal hydride material ages.

Once hydrogen is produced and stored, it must still be processed to perform useful work. Hydrogen is suitable for combustion in modified internal combustion engines. BMW has chosen this pathway for the implementation of hydrogen-powered vehicles in Germany. However, this reduces hydrogen to nothing more than an alternative fuel in combustion technology. A more elegant, though less mature, technology involves the direct conversion of the stored chemical energy to electrical energy in hydrogen fuel cells.

The fuel cell combines reduction and oxidation reactions to produce electricity directly from hydrogen without any need for mechanical or combustion processes as intermediate energy conversions. Very high conversion efficiencies can be obtained, yielding overall system efficiencies roughly double those that can be achieved with conventional energy currencies delivering equivalent services (Carette et al., 2001). Fuel-cell technology has become the subject of intense research, primarily for mobile applications, dominated by low-temperature devices based on proton exchange membrane (PEM) structures. Other low-temperature fuel cells include those based on alkaline electrolytes, or those that process methanol directly without using hydrogen in its free state. For stationary power applications, high-temperature fuel cells are being developed, in which the waste heat can be recovered and used to deliver heating services in concert with the electricity produced by the fuel cell itself. A good example is the cogeneration of household electricity and hot water or space heating.

Fuel-cell technology has seen spectacular progress over the past ten years, with huge improvements in performance and reliability, and similar decreases in costs. However, the technology remains expensive compared to any conventional generating alternative, primarily because of the expensive materials and manufacturing processes required. One area where fuel cells will soon be able to compete favourably with incumbent technologies is in low-power battery applications, such as for laptop computers and portable electronics. Although there is little short-term potential for climate change mitigation through deployment of fuel cells in these markets, they remain the most likely candidates for the early penetration of fuel cells in commercial settings and are therefore important for the long-term development of hydrogen technologies.

In each of the areas described (production, storage, and conversion), significant technological barriers must be overcome before a direct economic analysis of hydrogen technologies can be performed. However, hydrogen technologies break the link between energy sources and technological services, allowing all of our services to be delivered from an arbitrary (and hopefully carbon-free) mix of energy sources. By moving towards hydrogen, we also open up the possibility of the large-scale development of a carbon-free energy system. Without it, the hopes of the world for a better tomorrow rest on combustion-dependent oil-derived fuels.

Achieving Stabilization—How Much Carbon-Free Energy?

The world requires an immense amount of carbon-free energy in order to stabilize atmospheric concentrations of carbon dioxide (CO_2), and none of the current carbon-free alternatives are at a point where they can shoulder the burden currently carried by hydrocarbons. Hoffert et al., note that

"Researching, developing and commercializing carbon-free primary power technologies capable of 10–30 TW by the mid-twenty-first century could require efforts, perhaps international, pursued with the urgency of the Manhattan Project or the Apollo space programme" (1998). The scope of the problem is immense, prompting us to consider the sheer magnitude of the task of stabilizing or reducing atmospheric CO_2 concentrations. One common stabilization target is the doubling of pre-industrial atmospheric concentrations of CO_2 at 550 ppmv by 2100. The IPCC Working Group III states in its Summary for Policymakers, "known technological options could achieve a broad range of atmospheric CO_2 stabilization levels, such as 550 ppmv, 450 ppmv or below over the next 100 years or more" (IPCC, 2001).

Basing their analysis on the six basic emissions scenarios developed in the IPCC's *Special Report on Emissions Scenarios* (IPCC, 2000), Green and Lightfoot calculate the carbon-free energy required for stabilization in each case, and offer an example of one possible outcome based on not-unreasonable assumptions: "Assuming a population growth rate that falls from the current 1.3% a year to zero by 2100, a continuance of the long-term growth rate in GDP per capita of 1.6%, and a long-term rate of decline in energy intensity of 1.1% per year, it will take an estimated 1100 EJ/yr+ (35 TW+) of carbon-free power to stabilize the atmosphere CO_2 concentration at twice the pre-industrial level by 2100. In other words, a *20-fold or more increase* in carbon-free energy (power) [over 1999 levels] will be needed" (Green and Lightfoot, 2002; *emphasis added*). Recall that the world used 402 EJ of energy in 1999 and only 58.4 EJ that was of carbon-free energy. Green and Lightfoot suggest that the amount of carbon-free energy required in 2100 will be almost three times the *total* energy used in 1999. What is more, this figure implies an annual rate of reduction in carbon intensity (carbon emissions per unit energy) of greater than 1.2% sustained over more than a century, a figure more than three times the historical long-term rate, and for which there is no precedent. Providing a different perspective on energy demand, Kruger calculates that an additional 387 nuclear plants will be required by 2050 to meet the electrical generation requirements of a US vehicle fleet running on electrolyzed hydrogen (Kruger, 2000). This figure assumes that renewable generation will increase twenty-fold, and that nuclear technology will not improve beyond today's level. Even so, the magnitude of the energy required is almost beyond comprehension.

Conclusions

Climate change is above all the result of a century and a half of industrialization, and while we might like to find ways of mitigating climate change that do not involve "technological fixes," the sheer scope of the growing global dependence on energy-related services suggests that this is a false hope. Although there appear to be abundant stores of energy available from different sources that do not involve carbon cycles, none of these provides anything like the convenience and flexibility of hydrocarbons: displacing fossil fuels is going to be *hard.*

Although many place their hopes on renewable energy technologies, these technologies are still in their infancy and supply only a very small fraction of the overall energy mix today. While we might optimistically view this small fractional penetration as an opportunity for growth (which it is), there are simply too many technical and economic barriers facing renewables to conclude that their large-scale exploitation can provide the definitive solution to the problem of climate change. Nuclear power, though currently out of favour due to safety and waste-management concerns, presents the best opportunity for continuing today's high standard of living and for the expansion of these to the poorer people of the world. If managed correctly, nuclear and renewable generation—far from being the bitter enemies they are presumed to be today—could complement each other in interesting and productive ways.

Regardless of which mix of carbon-free technologies is chosen, the necessity for energy storage—particularly for transportation applications—will give rise to the development of a new energy currency that can complement electricity and in doing so provide the complete range of energy services currently delivered by hydrocarbons. This is the primary motivation for the development of hydrogen technologies that will allow transportation fuel to be produced from *any* energy source, including nuclear and renewables. Hydrogen is the only currency that provides complete separation of an energy source from the range of services it can deliver, and it therefore holds the key to a clean energy future.

The path to clean energy is still fraught with difficulties. The technologies involved are new and therefore comparatively expensive. The magnitude of energy required is daunting. Rather than an instant switch to these new technologies, it is more likely we will witness a continuation of the trend towards lower carbon fuels, expanded electrical services (particularly in transportation), and the eventual implementation of hydrogen generation and use in niche markets. Eventually, as hydrogen technologies evolve to the point where they become more functional than the incumbents, they will replace them.

References

Carette, L., K. Friedrich, and U. Stimming. (2001). Fuel Cells: Fundamentals and Applications. *Fuel Cells* 1(1): 5–38.

EIA. (2002). *International Energy Outlook 2002*. Washington, DC: US Department of Energy, Energy Information Administration, Office of Integrated Analysis and Forecasting.

Green, C., and H.D. Lightfoot. (2002). Achieving CO_2 Stabilization: An Assessment of Some Claims Made by Working Group III of the Intergovernmental Panel on Climate Change. Centre for Climate and Global Change Research, Report 2002-1. Montreal, QC: McGill University.

Herring, H. (1999). Does Energy Efficiency Save Energy? The Debate and Its Consequences. *Applied Energy* 63C3: 209–26.

Hoffert, M.I., K. Caldeira, et al., (1998). Energy Implications of Future Stabilization of Atmospheric CO_2 Content. *Nature* 395: 881–84.

IEA. (2000). *World Energy Outlook 2000*. Paris: International Energy Agency/OECD.

IPCC. (2000). *Emissions Scenarios*. Special Report of Working Group III of the Intergovernmental Panel on Climate Change. Cambridge, UK: Cambridge University Press.

———. (2001). *Climate Change 2001: Mitigation*. Cambridge, UK: Cambridge University Press.

Kruger, P. (2000). Electric Power Requirement in the United States for Large-Scale Production of Hydrogen Fuel." *International Journal of Hydrogen Energy* 25(11): 1023–33.

Laughton, M., and P. Spare. (2001). Limits to Renewables: How Electricity Grid Issues May Constrain the Growth of Distributed Generation. Energy World: *Journal of the Institute of Energy* (Nov. 2001): 8–11.

LNtV. (2002). Unpredictable Wind Energy: The Danish Dilemma, Landsforeningen Naboer til Vindmøller. <http://www.naboertilvindmoller.dk/English/danishdilemma.doc>

Mann, M.K., P.L. Spath, and W.A. Amos. (1998). Technoeconomic Analysis of Different Options for the Production of Hydrogen from Sunlight, Wind, and Biomass. In *Proceedings of the 1998 US DOE Hydrogen Program Review*, 28–30 April 1998, Alexandria, Virginia, vol. 1, 367–90. Golden, CO: National Renewable Energy Laboratory.

Morrison, R., D. Layzell, and G. McLean. (2002). Technology and Climate Change. *Canadian Nuclear Bulletin* 23(1).

Nakicenovic, N. (1996). Freeing Energy from Carbon. *Daedalus* 125(3): 95–112.

Niet, T.A. (2001). Modelling Renewable Energy at Race Rocks. *Mechanical Engineering*, MASc. Victoria, BC: University of Victoria.

Pettersson, L., and R. Westerholm. (2001). State of the Art of Multi-Fuel Reformers for Fuel Cell Vehicles: Problem Identification and Research Needs. *International Journal of Hydrogen Energy* 26(3): 243–64.

Rhodes, R. (1993). *Nuclear Renewal: Common Sense About Energy.* New York, NY: Whittle Books.

Smil, V. (1984). On Energy and Land. *American Scientist* 72(1): 15–21.

Turner, J.A. (1999). A Realizable Renewable Energy Future. *Science* 285: 687–89.

UNDP. (2001). Human Development Report 2001. New York: United Nations Development Programme.

UNSCEAR. (2000). Sources and Effects of Ionizing Radiation : United Nations Scientific Committee on the Effects of Atomic Radiation. New York: United Nations Scientific Committee on the Effects of Atomic Radiation.

USDOE. (1999). Clean Coal Technology: The Investment Pays Off. Washington, DC: US Department of Energy.

Webster, C. (2002). Compressed Hydrogen Storage Systems for Automotive Vehicles. *Proceedings of the World Hydrogen Energy Conference*, Montreal, 14–17 June 2002.

Whitlock, J. (2002). Weekly Electrical Production by Canadian Nuclear Reactors. Canadian Nuclear FAQ. <http://www.nuclearfaq.ca/nuke-gen-weekly.htm>

Economic Aspects of Climate Change

G. Cornelis van Kooten

As DISCUSSED IN CHAPTERS 2 AND 3, climate change is a long-term threat to the earth's ecosystems and to the way people lead their lives. Climate change constitutes a threat to agriculture, particularly subsistence farming in developing countries, and to coastal dwellers who could lose their homes as a result of flooding caused by rising sea levels. While the full extent of the potential damages from global warming remains unknown, scientists have argued that action should be taken to mitigate potentially adverse consequences. However, in such arguments, economic considerations have often been neglected. This chapter therefore examines the economic aspects of climate change mitigation and adaptation.

Although other greenhouse gases (GHGs) are important, the focus is on carbon dioxide (CO_2) because, as Weaver notes in chapter 2, it is the most important driver of an enhanced anthropogenic greenhouse effect. There are two principal means for reducing atmospheric CO_2: reducing emissions of CO_2 from fossil-fuel burning and removing CO_2 from the atmosphere by sequestering carbon in terrestrial ecosystems (e.g., increasing wood biomass and soil carbon). Although equivalent in their effect on climate change, emissions reductions and terrestrial carbon uptake pose substantially different challenges for policy-makers.

Economic Instruments for Mitigating Climate Change

The economic objective of climate change policy is to reduce GHG emissions at the lowest possible cost. A number of regulatory and market instruments are available (Stavins, 2002). Governments have traditionally favoured command-and-control regulations to address environmental spillovers. Regulations accomplish environmental targets by setting uniform standards that specify either the means to be used in order to comply with a regulation or a uniform control target, allowing some latitude in how the target is achieved. The regulatory approach is generally expensive because it fails to acknowledge differences among firms and provides no incentives to seek least-cost options for achieving targets. One alternative to regulation has been voluntary compliance, but this has resulted in only limited success and cannot be relied upon to achieve the reductions in CO_2 emissions required to meet Kyoto Protocol targets (see Takahashi et al., 2001). As a result, governments have increasingly turned to market instruments, with emissions trading seemingly becoming the instrument of choice.

In emissions trading, the authority sets a cap on emissions and then allocates permits to emitters according to their historic emissions (known as "grandfathering") or their willingness to purchase auctioned permits. Once permits are allocated, firms are free to buy and sell them. Hence, this instrument is referred to as a cap-and-trade scheme. When permits are auctioned, the revenue can be used to reduce extant distortions in the tax system ("recycled"), thereby lowering overall costs. The problem with a cap-and-trade scheme is that transaction costs increase when terrestrial carbon offsets are included (discussed below), and it has been shown that a cap-and-trade scheme is costlier than a tax that achieves the same level of emissions reduction (Pizer, 1997). However, in many cases cap-and-trade schemes are "optimal" simply because they are the only politically acceptable alternative to command-and-control regulations (Stavins, 2002).

The other market instrument is a carbon tax/subsidy scheme. Activities that release carbon to the atmosphere, whether the result of fossil-fuel burning, cultivating grasslands, or felling trees, are taxed according to the amounts of carbon they release, while activities that remove carbon from the atmosphere (e.g., planting trees or converting land from cultivation to pasture) are subsidized in like manner (see van Kooten et al., 1995). If revenue recycling occurs, a carbon tax/subsidy is considered the least-cost means to achieve Kyoto targets. Further, as long as carbon-equivalent emissions (including those from land use, land-use change, and forestry [LULUCF] activities) exceed carbon uptake, tax revenues can be used to fund the subsidy side of the scheme.

Kyoto Protocol Instruments

Policy implementation is complicated by the fact that the Kyoto Protocol permits an Annex B country to achieve its CO_2 reduction target in different ways:

- Countries can simply reduce their own emissions of GHGS to the target level.
- Rather than reducing domestic CO_2 emissions to the target level, a country can achieve its target by sequestering an equivalent amount of carbon in domestic terrestrial ecosystems. Under Kyoto Protocol Article 3.7, permitted activities are those that reduce carbon release from deforestation in Annex B countries that had net LULUCF emissions in 1990; net removals by sinks resulting from human-induced afforestation, reforestation, and deforestation (ARD) (Article 3.3);[1] and net removals by changes in agronomic practices (cropland and grazing land management, and revegetation actions) and enhanced forest management (Article 3.4).
- Joint implementation is encouraged under Article 6. Joint implementation allows an Annex B country to participate in emissions reduction or carbon sequestration in another Annex B country, thereby earning "emission reduction units" (ERUS) that are credited toward the country's own commitment.
- Under the "clean development mechanism" of Article 12, an Annex B country can earn "certified emission reductions" (CERS) by funding an emissions reduction or carbon sequestration project in a non-Annex B (developing) country. However, only afforestation and reforestation activities can be used to generate carbon uptake CERS, and their use is limited (in each year of the commitment period) to 1% of the Annex B country's 1990 (base-year) emissions.
- Finally, an Annex B country can simply purchase excess emission permits from another Annex B country (Article 17). Emission permits in excess of what a country needs to achieve its commitment are referred to as "assigned amount units" (AAUS) that can be purchased by other countries. These are particularly important to economies in transition that easily attain their KP targets because of economic contraction and slow growth and the concomitant closure of inefficient power plants and manufacturing facilities; the resulting AAUS—"hot air"—may be sold at whatever price is available.

The purchase and sale of AAUS, CERS, and ERUS is referred to as "international emissions trading."

In principle, enabling emitters to purchase carbon offsets (or carbon sink credits) in lieu of emission permits, and allowing signatories to purchase such credits (offsets) from developing countries via the CDM, should reduce compliance costs because it should not matter how CO_2 is removed from the atmosphere. When CO_2 emissions are the sole concern, the government only has to set the emissions cap and then allocate the required number of permits. With the addition of carbon offsets and other methods for achieving emission reductions, the authority needs more information than it previously did. In order to set the cap on emissions, it needs to know something about the potential supply of terrestrial carbon offsets in each future year, plus the supply of carbon offsets from joint implementation and the clean development mechanism. Integrating a market for carbon credits into a market for emission permits increases transaction costs. As well, several practical questions related to the integration of carbon sinks in a cap-and-trade scheme remain.

The Complexity of Carbon Sinks

Clearly, a market-based solution will be effective only in the presence of certain market conditions. For example, emission reductions and carbon offsets need to be certified, and a method for making exchanges between emission reductions and offsets needs to be found; as well, an overseeing body with well-defined rules and regulations is needed. Even if all conditions for trade are present, there remains concern about the extent of agriculture's willingness to participate in GHG trading programs due to other indirect costs or barriers to entry, such as uncertainty about farm payments and subsidies, implications for trade, or transaction costs associated with the creation of carbon sinks on agricultural land (van Kooten et al., 2002).

It is imperative to identify methods that will encourage both agricultural landowners' willingness to create carbon credits for trading and also their capacity to create and market carbon offsets. In suggesting the use of sinks as a flexible mechanism for meeting CO_2 emissions goals, it is important to understand landowners' incentives, motivations, and preferences in order to create a well-functioning market. Landowners' preferences for different carbon sequestration methods are likely influenced by the available information and methods, institutional support and structure, and relative risk and uncertainty with regards to maintaining a profitable enterprise and remaining eligible for government programs.

Recent research by van Kooten et al. (2002) into landowners' willingness to plant trees or otherwise create carbon credits suggests that costs may be higher than is often thought, particularly in previously treed regions where owners or their forebears incurred substantial sacrifice to carve out farms.[2] Further, landowners appear reluctant to engage in carbon offset trading, stating that they prefer contracts with governments, NGOs, or even large firms to conduct specified LULUCF activities. That is, landowners prefer to enter into a contract to modify their land use rather than to sell carbon credits per se. The problem is that contracting may have associated transaction costs that could significantly increase costs of creating carbon credits (Suchánek et al., 2002; van Kooten, 2000).

The other problem of mixed emission-offset carbon trading relates to the conversion factor. Compared to not emitting CO_2 from a fossil-fuel source, terrestrial sequestration of carbon is unlikely to be permanent, particularly for carbon stored in agricultural sinks. However, temporary removal of carbon is important because it postpones climate change, buys time to improve technology and replace fuel-inefficient capital equipment, may cost less than simply reducing CO_2 emissions, and allows time for learning. In addition, there is the possibility that some temporary sequestration may become permanent (Marland et al., 2001, p. 262). The ephemeral nature of terrestrial carbon uptake can be addressed by providing partial credits for stored carbon according to the perceived risk that it will be released at some future date, or some assurance that the temporary activity will be followed by one that results in a permanent emissions reduction. Alternatively, a conversion factor that translates years of temporary carbon storage into a permanent equivalent can be specified.

The IPCC (2000) uses the notion of ton-years to make the conversion from temporary to permanent storage. Suppose that one-ton of carbon-equivalent GHG emissions is to be compensated for by a ton of permanent carbon uptake. If the conversion rate between ton-years of (temporary) carbon sequestration and permanent tons of carbon emissions reductions is k, a LULUFC project that yields one ton of carbon uptake in the current year generates only $1/k$ tons of emission reduction—that is, to cover the one-ton reduction in emissions requires k tons of carbon to be sequestered for one year. The conversion rate ranges from 42 to 150 ton-years of temporary storage to cover one permanent ton, with central estimates around 50:1.

As Marland et al. (2001) point out, the ton-year accounting system is flawed: ton-year credits (convertible to permanent tons) can be accumulated while trees grow, for example, with an additional credit earned if the biomass is subsequently burned in place of an energy-equivalent amount of fossil fuel (p. 266). To avoid such double counting and the need to establish a conversion factor, the authors propose a rental system for sequestered carbon. A one-ton emission offset credit is earned when the sequestered carbon is rented from a landowner, but, upon release, a debit occurs: "Credit is leased for a finite term, during which someone else accepts responsibility for emissions, and at the end of that term the renter will incur a debit unless the carbon remains sequestered *and* the lease is renewed" (p. 265, emphasis in original). In addition to avoiding the potential for double counting, the landowner (or host country) would not be responsible for the liability after the (short-term) lease expires. Further, rather than the authority establishing a conversion factor, the market for emission permits and carbon credits can be relied upon to determine the conversion rate between permanent and temporary removals of CO_2 from the atmosphere. A carbon tax/subsidy scheme would accomplish the same thing.

Costs and Benefits of Mitigation

In analyzing environmental policies, economists always ask whether society is better off with the policy in place or without it. Cost-benefit analysis (CBA) employs this principle using a money metric. The use of this metric raises substantial ethical and economic dilemmas. The purpose in this section is not to delve into these issues, which have been previously addressed, but rather to provide information on the costs of mitigating climate change, the potential benefits of so doing, and, finally, whether mitigation or adaptation is the economically preferred strategy, at least given current economic knowledge.

Global costs of mitigating climate change

Three modelling approaches are used to estimate the costs of reducing CO_2 emissions:

- Energy-technology models are bottom-up in the sense that they examine the technical feasibility of implementing emissions reduction projects without regard for how these affect the economy as a whole. Such models generally include the maximization of some objective function over multiple periods with between-period constraints.
- Macroeconomic or econometric models use statistical (historical) data for estimating economic and technological relationships. Scenarios developed by energy-technology models can be used to drive macroeconomic (econometric) models, but they lack constraints that might

limit certain activities, and they fail to account for rising prices. For example, the model may have no constraint on the availability of trained labour even though constraints in the real world may cause wages for trained labour to rise dramatically.

- Finally, computable general equilibrium (CGE) models are top-down in nature as they explicitly account for all constraints and interactions within the economy. Because CGE models assume that economic resources are allocated efficiently, they do not permit mitigation strategies at negative costs (as this would imply an inefficient initial allocation of resources).

Estimates of the marginal costs of abating CO_2 emissions to meet 2010 KP targets from a variety of energy-technology, macroeconomic, and CGE models are reported in Table 7.1. In these models, carbon emission permits are used as the instrument for mitigation, leading to an implicit carbon tax for achieving targets, and revenues are recycled in lump-sum fashion rather than used to reduce distortionary taxes elsewhere in the economy; no environmental co-benefits are included (Ekins and Barker, 2001). Models differ in their outcomes because of differences in how coefficients are parameterized and because they focus on different regions and/or sectors.

US domestic marginal abatement costs range from $76/t C to $410/t C in the absence of emissions trading, while costs in other regions are estimated to be higher, in some cases (Japan) much higher. If trading of permits is allowed among Annex B countries only, costs fall to $20–$135/t C, while global trading would reduce them even further—down to $5–$86/t C. If tax revenues had been used to reduce distortions elsewhere, the cost estimates in Table 7.1 would have been somewhat lower yet. Average marginal abatement costs fall from $186–$303/t C (depending on the region) to $78/t C if permit trading occurs among Annex B countries and down to $29/t C with full global trading.

More detail about the potential costs of meeting the Kyoto targets is provided in Table 7.2. Projected reductions in GDP in 2010 are provided for four global regions. Average loss in GDP is lowest for Japan (about 0.8%) and Europe (nearly 1%), and highest for Canada, Australia, and New Zealand (CANZ) at 1.5%, followed by the US at 1.3%. For the US, this amounts to more than $120 billion annually; for Canada, Australia, and New Zealand, it amounts to some $17 billion per year. These are not insignificant costs. And they continue in perpetuity. Of course, emissions trading has the potential to reduce costs significantly.

The results in Tables 7.1 and 7.2 are idealistic in the sense that they assume no policy failure (except that revenues are recycled in lump-sum fashion rather than used to reduce distortionary taxes elsewhere in the economy). However, in countries where CO_2 taxes have been introduced, some sectors have been exempted or taxed at a different rate, thereby substantially increasing economic costs relative to a policy that involves uniform taxes (IPCC, 2001, p. 56).

Table 7.1 Energy Modeling Forum Main Results: Marginal Abatement Costs to Achieve 2010 Kyoto Target (in 1990 US$/t C)[a]

	No Trading				Trading Only	
Model	**USA**	**OECD Europe**	**Japan**	**Canada, Australia New Zealand[b]**	**among Annex B Countries[b]**	**Full Global Trading[b]**
Energy-Technology Models						
Administration	154	43	18	n.a.	n.a.	n.a.
EIA	251	110	57	n.a.	n.a.	n.a.
CETA	168	46	26	n.a.	n.a.	n.a.
Funds	14	10	n.a.	n.a.	n.a.	n.a.
GRAPE	204	304	70	44	n.a.	n.a.
MERGE3	264	218	500	250	135	86
RICE	132	159	251	145	62	18
POLES	136	135	195	131	53	18
Econometric Models						
Oxford	410	966	1074	224	123	n.a.
CGE Models						
ABARE-GTEM	322	665	645	425	106	23
AIM	153	198	234	147	65	38
G-Cubed	76	227	97	157	53	20
MIT-EPPA	193	276	501	247	76	n.a.
MS-MRT	236	179	402	213	77	27
SGM	188	407	357	201	84	22
WorldScan	85	20	122	46	20	5
Average	**187**	**248**	**303**	**186**	**78**	**29**

Source: IPCC, 2001, Table TS.4, and calculation.

Notes: [a] To convert to $ per t CO_2, multiply the figures in the above table by 12/44 as 1 t C = 44/12 t CO_2.
[b] n.a. indicates not available.

Table 7.2 Reduction in GDP in 2010 as a Result of Meeting Kyoto Protocol Targets (% of GDP)

	No Trading				Annex B Trading				Full Global Trading			
Model	**USA**	**OECD-Eur**	**Japan**	**CANZ**	**USA**	**OECD-Eur**	**Japan**	**CANZ**	**USA**	**OECD-Eur**	**Japan**	**CANZ**
ABARE-GTEM	1.96	0.94	0.72	1.96	0.47	0.13	0.05	0.23	0.09	0.03	0.01	0.04
AIM	0.45	0.31	0.25	0.59	0.31	0.17	0.13	0.36	0.2	0.08	0.01	0.35
CETA	1.93	n.a.	n.a.	n.a.	0.67	n.a.	n.a.	n.a.	0.43	n.a.	n.a.	n.a.
G-Cubed	0.42	1.5	0.57	1.83	0.24	0.61	0.45	0.72	0.06	0.26	0.14	0.32
GRAPE	n.a.	0.81	0.19	n.a.	n.a.	0.81	0.1	n.a.	n.a.	0.54	0.05	n.a.
MERGE3	1.06	0.99	0.8	2.02	0.51	0.47	0.19	1.14	0.2	0.2	0.01	0.67
MS-MRT	1.88	0.63	1.2	1.83	0.91	0.13	0.22	0.88	0.29	0.03	0.02	0.32
Oxford	1.78	2.08	1.88	n.a.	1.03	0.73	0.52	n.a.	0.66	0.47	0.33	n.a.
RICE	0.94	0.55	0.78	0.96	0.56	0.28	0.3	0.54	0.19	0.09	0.09	0.19
Average	**1.30**	**0.98**	**0.80**	**1.53**	**0.59**	**0.42**	**0.25**	**0.65**	**0.27**	**0.21**	**0.08**	**0.32**

Source: IPCC 2001, Table TS.5, and calculation.
Note: n.a. = not available.

Mitigation cost estimates for Canada

In 1990, Canada generated 607 Mt of CO_2-equivalent emissions (165.5 Mt C); thus Canada's Kyoto commitment requires a reduction of emissions to 571 Mt CO_2 (155.7 Mt C) by 2008–2012. By 2000, after a period of economic expansion, Canada's GHG emissions had increased by 19.6% to 706 Mt CO_2 (192.5 Mt C) (Olsen et al., 2002). Business-as-usual emissions are projected to reach 802 Mt CO_2 (218.7 Mt C) annually by 2010, some 40% above Canada's Kyoto commitment to reduce emissions 6% below 1990 levels by 2008–12. Canada must reduce emissions by 240 Mt CO_2 per year below business-as-usual emissions. Several studies have sought to address the costs of complying with Kyoto targets. The main results of these studies are summarized in Table 7.3, but these hide a fair amount of detail.

Table 7.3 Summary of Costs of Attaining Kyoto Protocol Targets at 2010, Canada

Item	No Emissions Trading	Global Emissions Trading
Required Carbon Tax (US$/t c)	150–835	11–114
Welfare loss (% of GDP)	0.9–2.2	0.2–0.6

Source: Wigle, 2001, pp. 6–7.
Note: Not including studies by AMG (2000) and Wigle (2001).

For Canada, cost estimates from two energy-technology models indicate that, at the margin, economy-wide abatement costs are likely to be between C$57 and C$120 per tonne of CO_2 if Canada acts alone (Analysis and Modelling Group, National Climate Change Program, [hereafter AMG], 2000). This suggests costs of C$209–440 per t C (as one t C translates into 3.667 t CO_2), which might be considered on the high side. Estimates of the GDP loss in 2010 derived from such bottom-up models range from 0–0.5% of GDP, with some technologically optimistic models even suggesting gains in GDP (Wigle, 2001).

Results from one macroeconomic modelling exercise that uses outputs from two energy-technology models indicate that, for Canada to attain KP targets, GDP could be reduced anywhere from 1% to 3% in the short term (to 2012) and by 4% in the period beyond 2012 (AMG, 2000). To provide a perspective on these estimates, "a reduction in GDP of 3 percent in 2010 means that, over the decade [2000–2010], the economy will grow by about 26 percent instead of 30 percent as projected in the reference case. This is equivalent to the loss of roughly one year's growth, or, ... in 2010, the loss in annual economic output of approximately $40 billion (or $1,100 per capita)" (AMG, 2000, p. xiv). In his review of results for Canada from macroeconomic models, Wigle (2001) finds they give GDP reductions in 2010 ranging from 0.4 to 2.3%.

Unlike macroeconomic models, output from bottom-up models cannot be directly inputted into a CGE model. Rather, policies to attain the Kyoto target are modelled via taxes or permit trading. Assuming lump-sum transfers back to citizens so that the total tax burden is unaffected, AMG (2000) results indicate that Canadian GDP would fall by 0.3% (if global emissions trading is allowed and permits are priced at no more than C$24/t C) to 0.9% of GDP (if Canada acts alone). Using a CGE model based on data from the Global Trade Analysis Project (which allows for international emissions trading), Wigle (2001) finds that a cost-effective domestic implementation scheme would reduce well being by about 1.3% of GDP, or about C$13 billion annually. Finance Canada simulated industry-specific policies identified by the Issues Tables using a CGE model: if targeted measures could be deployed at relatively low cost, GDP would still fall by an estimated 0.9% ($9 billion annually) in attaining the Kyoto target (McKitrick and Wigle 2002, p. 23).

Most models assume that governments implement the best policies for achieving Kyoto Protocol targets, which is generally not the case. For example, since the energy-intensive industrial sectors are likely to be big losers from carbon taxes (emission permits) and competitiveness of other strategic sectors might be jeopardized, most countries have provided certain vulnerable and/or strategic sectors with tax exemptions or concessions (Ekins and Barker, 2001, pp. 354, 369). It is likely that sector concessions will characterize any Canadian policy, simply because regional considerations may be of more importance than whether optimal instruments are employed, although the US decision to abandon the Kyoto process will also play a role in sector targeting. Wigle (2001) finds that, by exempting the non-energy intensive sectors of the economy, costs increase by about 0.5% of GDP, but, by exempting the energy-intensive sectors, GDP will be reduced by 7.5%, or about C$75 billion per year. Rather than exempting energy-intensive sectors from a carbon tax, it might be preferable to pay back some of the tax revenue to those sectors in lump-sum fashion.

The importance of choosing the "correct" policy instrument cannot be overlooked. Simulations using the TIM (The Informetrica Model) econometric model indicate that, if the government distributes emissions permits freely rather than auctioning them off, the cost of complying with the KP may be some $1,100 to $3,000 per capita higher in 2010 (McKitrick and Wigle, 2002, p. 22). In analyzing the government's implementation plan, van Kooten (2003) argues that Canada will fail to attain its Kyoto Protocol targets and that what is attainable will annually cost $1.4–$3.0 billion more than need be the case, because of failure to rely on markets to a greater extent.

Some general conclusions about the costs of mitigating global warming can be drawn. First, market instruments such as taxes and emission permits will lead to lower costs of achieving emissions targets than will regulation. Second, carbon taxes or emissions trading are likely to result in adverse distributional consequences, both in terms of their effect on those in low-income groups (see Ekins and Barker, 2001) and on the energy-intensive industrial sector. Third, many of the local environmental co-benefits of reducing CO_2 emissions have been overlooked in calculating the costs of meeting targets, although their magnitude is likely small compared to costs (Kennedy, 2002). Fourth, opportunities exist for reducing CO_2 emissions at zero or negative cost, but these are not always adopted because of market and/or government failure (e.g., government agricultural subsidies, tax breaks for production of coal and steel not available to other sectors, trade protection against imports of steel and wood products). Finally, economic models are still in their infancy, and some believe that these models probably overstate the costs of achieving Kyoto Protocol targets, although "this is a matter of some controversy" (Ekins and Barker, 2001, p. 369).

Potential benefits of mitigating global warming

Perhaps the most challenging area of economic research concerns the estimation of the potential damages of climate change. First, given scientific uncertainties, it will not be clear whether a particular future event, such as a drought, an unusual storm, or an early spring (which negatively affects tree growth in boreal climes), can be attributed to climate change of anthropogenic origin, or whether it is simply a natural occurrence well within the usual vagaries of weather patterns. In the future, it will probably be a bit of both; but the important question is, what would the subsequent weather-related damage of this event be if the Kyoto Protocol were completely implemented versus what it would be without Kyoto? Even if this question can be answered with a reasonable degree of certainty, two other hurdles remain: future events of this nature need to be forecast with some probability, and estimates of expected damages need to be provided (again with some probability). Clearly, estimating the benefits to be gained by mitigating climate change (i.e., the damages avoided) is a much more difficult and uncertain prospect than estimating costs. Since costs are related to what is happening in the economy today, they are easier to measure. Furthermore, more effort has already been devoted to measuring costs than to estimating potential damages. However, as noted above, even estimates of the costs of achieving Kyoto Protocol targets are controversial; how much more controversial, therefore, will be estimates of the damages avoided.

A number of studies have attempted to estimate the damages that climate change might cause; avoiding these would then be the benefit of mitigating climate change. It is important to recognize that extant damage estimates are not to be equated to the benefits of attaining Kyoto targets; the Kyoto Protocol itself will likely do little to prevent global warming (see below) so not all related damages would be prevented even if targets were met. Nor do all authors come to the conclusion that climate change is only "bad" in terms of its impact on global well being, although all studies agree that some regions will lose while others will gain from an increase in global temperatures. In this regard, it is important to note that, even if climate change results in benefits, but with some probability, a risk adverse society may still be willing to pay some amount to avoid climate change.

Darwin et al. (1995) use a CGE model to estimate the global welfare impacts of climate change as it affected output in the primary sectors. CO_2-fertilization benefits that would cause crops and trees to grow faster were ignored.[3] The study found that, if landowners were able to adapt their land uses to maximize net returns, global GDP would increase by 0.2–1.2% depending on the GCM projections employed. Mendelsohn et al. (1994)

project an increase in US GDP as a result of global warming. Moore (1998) estimates that a global temperature increase of 4.5°C will yield some \$30–\$100 billion in health benefits to US residents. Given that Arizona and Nevada have been the fastest growing states in the US, it does appear that people express a preference for living in warm (even hot) and dry climates. Empirical measures of the values of these amenities are generally lacking. Lise and Tol (2001) also found that people have a preference for warmer climates, as evidenced by their choices regarding vacation destinations. In contrast, results from Maddison and Bigano (2000) are less conclusive: in Italy, they found that higher summer temperatures are regarded negatively, but so are lower January temperatures and higher January precipitation.

Most authors take the view that, when health, sea-level rise, and other impacts are considered, climate change will result in overall global damage. Heal and Kriström (2002) report that Horowitz (2001) used a cross-section of 156 countries to examine the relationship between temperature and income per head, concluding that a 1% increase in temperature (measured in °C) leads to a 0.9% decrease in income, which implies that a 3°C increase in temperatures will reduce global GDP by 4.6%. This is only suggestive as there is no causal mechanism and precipitation is ignored, as are amenity values.

Many studies simply assume a relationship between damages and temperature change. For example, Nordhaus (1994) and Nordhaus and Boyer (2000) model damages from global warming using the following equation:

$$D = c \times \Delta T^2 \times Y,$$

where ΔT is the change in temperature (in °C), Y is global output (about US\$28 trillion in 2000), and c is a proportionality constant (= 0.0133/3°C) that says that, for ΔT = 3°C, global output falls by 1.33% (compared to the 4.6% decline noted above). Emissions per unit of output in 2000 were 0.37 Gt C per \$1 trillion of output, and are expected to fall to 0.27 Gt C per \$1 trillion by 2050 (see Newell and Pizer, 2001, p. 4). For a projected 2°C rise in temperatures, this equation suggests that damages could amount to nearly \$250 billion per year, although this scenario is not likely to be realized until mid-to-late century.

Other relationships between global temperature rise and GDP can be identified, but any such equation is clearly speculative. With the exception of indirect statistical analyses (see, for example, Horowitz, 2001), empirical estimates of global damages avoided are unlikely to be available for quite some time. Probably the best present option is to obtain damage estimates from models that draw from GCM and other biophysical models, and couple the estimates onto a global CGE or econometric model, as Darwin et al. (1995) and Mendelsohn et al. (1994) have done. However, the final estimates of global damages (or benefits) from such an exercise would have a high degree of uncertainty, and would certainly be controversial.

Comparing costs and benefits

Given the high degree of uncertainty, and even speculation, about both the costs of mitigating climate change and the damages avoided by mitigation, what can be said about the balancing of costs and benefits—about an optimal policy response? A number of issues need to be considered, beginning with the question of mitigation versus adaptation.

Mitigation vs. Adaptation

Suppose that all signatories to the Kyoto Protocol, including the US, achieve their targeted emissions reductions. Even in that unlikely event, the impact on the concentration of atmospheric CO_2 will be small. Currently, the level of CO_2 in the atmosphere is 370 ppmv (parts per million by volume), with total carbon in the atmosphere of some 780 Gt.[4] Average annual emissions of carbon are 1.14 t C per person, so about 6.9 Gt C enters the atmosphere each year. The total annual reduction of GHG emissions required of Annex B countries amounts to a CO_2-equivalent of some 250 Mt C, which is less than 4% of annual emissions and only 0.03% of the atmospheric stock of carbon. Clearly, the Kyoto Protocol can do little more than delay warming by a few years.[5] In that case, its benefits are almost negligible.

Crudely put, if the costs of achieving Kyoto Protocol targets are high, it might simply be cheaper to adapt to climate change rather than prevent it from happening. One approach is to put aside the money that would otherwise have been spent on mitigation in a special fund to compensate those who could, at any date in the future, demonstrate that they have been injured as a result of climate change.

In 2000, the GDP of the regions in Table 7.2 amounted to $22,570 billion. Suppose that, rather than incurring costs to mitigate climate change, these countries annually invested 0.5% of their GDP (which is less than the annual cost of complying with Kyoto and much less than that required to stop atmospheric CO_2 from growing). Thus, countries would place $112.85 billion in a contingency fund this year, $116.24 billion next year if global GDP grew by 3%, and so on. Not including additional annual contributions, we might expect the contingency fund to grow at a rate of 7%, as this has been the historical average real rate of return to large companies over the period 1926–1990 and is the rate used by the US Office of Management and Budget for standard cost-benefit analysis (Newell and Pizer, 2001, p. 6). As long as current *identifiable* damages from global warming are below about $112 billion and rise at a rate less than or equal to the rate of return (7%) plus the rate of growth of GDP, the contingency fund will always be adequate to cover damages (see McKitrick, 2001).

The highest damage estimate provided in the previous section was 4.6% of global GDP, but that is only reached after average global temperatures have risen by 3°C over the current mean. This is projected to occur in 50–100 years. Suppose it happens in 50 years, and damages rise at a constant rate between now and then. Then, using a numerical simulation model, current damages of less than $1 million (easily covered by the contingency fund) would need to grow at less than 1.7% per year to achieve damages of 4.6% of global GDP 50 years from now. Thus, as long as the rate of return to contingency funds exceeds 1.7%, it will be possible to cover quite easily any future damages from climate change.[6]

Several factors have been ignored in this analysis. First, by mitigating now, local environmental co-benefits of reducing fossil-fuel pollution can be realized. Second, the transaction costs of operating an invest-and-compensate scheme (including identifying when compensable damages have occurred and the making of payments) might be substantial. Third, and perhaps the biggest obstacle to implementation of a compensation scheme, will be the inability of governments to avoid appropriating contingency funds for other purposes, particularly as the fund increases to become a substantial size. Finally, there may well be complementarities between mitigation and adaptation strategies that should not be overlooked in judging adaptation to be preferred (see Kane and Shogren, 2000).

Early Action

In the context of the Kyoto Protocol, Kennedy (2002) asks whether taking early action on mitigation (an emissions tax, credit for early action, or an early cap-and-trade program) is optimal. He identifies two main benefits of early action: potential compliance cost savings during 2008–2012 (from early investment) and early environmental benefits. He also points out that early action is not the same as early emission reductions, because any early reduction in emissions is not credited to a country under the Kyoto Protocol. Hence, he distinguishes between technical abatement and behavioural abatement. The former results from technical improvements and leads to a permanent reduction in emissions, while the latter often results in ephemeral abatement (e.g., taxes affect behaviour in the desired direction, but emissions return to their former level once taxes are removed unless structural changes have occurred). Kennedy (2002) concludes that "policies that focus on early actual emission reductions will tend to distort abatement investment decisions and thereby inflate the national compliance cost of the Kyoto target" (p. 32–33). Credits for early action or an "aggressive early cap-and-trade program have the potential to be highly distorting [and] the associated welfare losses could be billions of dollars" (p. 33). Early action seems justified only if there are significant environmental co-benefits, but these are likely best addressed separately, rather than relying on instruments that target carbon.

Uncertainty

Analysis of climate change policy must take into account scientific uncertainty related to future climate regimes and the way that the altered climate will impact on people in terms of their physical environment, as well as uncertainty about changes in economic opportunities. In the presence of such uncertainty, there is a real economic benefit to avoiding irreversible decisions. By remaining flexible, society can take advantage of new information as time passes. In the context of climate change, there are two types of irreversibility to consider.

If GHG emissions are not abated, there could be an irreversible shift in climate regimes (such as the oft-cited "flip-flop" in the Gulf Stream). By mitigating climate change, society retains the flexibility to benefit from new information about the potential adverse (or even beneficial) consequences of climate change on the earth's ecosystems and its people. Retaining our options and avoiding climate irreversibility certainly has value.

Further, by *not* taking action now, society avoids irreversible investments in mitigation technology and leaves open opportunities to benefit from better methods for mitigating climate change or adapting to it in the future. This benefit of flexibility may be substantial and can justify not taking action to mitigate climate change, just as the previous argument can be used to justify current expenditures to mitigate global warming.

Which of these two values is the larger? The economics literature indicates that it hinges on people's prudence, which, in technical jargon, is determined by the third derivative of the utility function. In their review of the literature on decision-making under uncertainty, Heal and Kriström (2002) conclude that, in addition to knowledge about the probability distribution of the effects of climate change, risk attitudes (i.e., the utility function), growth rates, rates of technological advance, and discount rates are important. That is, economists cannot now provide a definitive answer to the question, because research on the economic front is lagging behind that in the physical sciences. Nonetheless, the decision is as much an economic one as anything else.

Discussion

The general conclusion from economic research is that market instruments will be more effective in reducing atmospheric concentrations of CO_2 than command-and-control regulation. While a carbon tax/subsidy scheme that recycles revenues in an appropriate manner is the least-cost means of achieving Kyoto targets, a grandfathered cap-and-trade scheme that includes carbon credit markets is likely to be favoured by policy-makers for political reasons. However, economic considerations also suggest that immediate action to prevent climate change through aggressive mitigation policies may be premature, although there may be room for some action, perhaps even action compatible with Kyoto requirements.

Unlike the results from general circulation models, which appear to be unequivocal about the direction of climate change, economic research has not achieved a level of consensus. Since the Kyoto Protocol is likely to have a relatively minor impact on atmospheric concentrations of CO_2, most economic models seem to favour adaptation over mitigation. In that case, much greater emphasis needs to be placed on adaptation strategies and on adaptation research, including economic research into the development of strategies for addressing periodic water shortages that impact dryland and irrigated agriculture, hydropower generation, and water for wildlife habitat.

What about no-regret strategies? Although economists generally eschew the notion that economic agents fail to take advantage of profitable opportunities that also reduce CO_2 emissions (e.g., adopting conservation tillage because of its supposedly higher net returns, purchasing longer-lasting fluorescent light bulbs), market failure and/or policy failure may prevent adoption of costless opportunities to reduce atmospheric CO_2. Governments subsidize coal mining, protect inefficient industries that are large energy users (e.g., recent protection of the US steel industry), encourage more energy-intensive agriculture, and employ other implicit and explicit subsidies that encourage land conversion and/or energy consumption. Lobbying has prevented the application of fuel efficiency standards to sport-utility vehicles and trucks in the US. Lobbying by US lumber producers to restrict Canadian imports of softwood lumber has increased prices, resulting in a shift towards the use of more energy-intensive building products (steel, aluminium and cement) and less carbon storage in wood products. Such policies and exercise of market power do much to encourage activities that enhance climate change.

One other no-regrets strategy has more to do with global politics than economic science. The US and other western countries are increasingly reliant on oil from the Middle East: as its own reserves dwindle, the proportion of US exports accounted for by oil from the Arabian Gulf increases. In the light of US energy security issues and the events of September 11, 2001, the US might want to make every effort to reduce its oil consumption by implementing higher consumption taxes, adopting stringent and more encompassing fuel efficiency standards, and taking other actions that reduce dependence on Gulf oil. From the perspective of climate change, these are appropriate no-regret strategies, but they need to be adopted for practical, political reasons.

Notes

1 Not included is the COP 6_{bis} (COP 7) provision that a country can offset in any year of the commitment period an accounting deficit under Article 3.3 (e.g., from clear cutting), with a net increase in sinks due to forest management under Article 3.4 to a maximum of 8.2 (9.0) Mt C.

2 Forestland continues to be cleared for agriculture. For the two-year period 1995–1997, for example, 0.7% of Alberta's forestland (some 200,000 ha) was converted to agriculture (Alberta Environmental Protection, 1998).

3 Research on the effects of increased CO_2 on plant growth suggests that these might be substantial (see Fround-Williams et al., 1996).

4 The conversion factor from ppmv to Gt C is 2.13 Gt C/ppmv (Newell and Pizer, 2001, p. 4).

5 If Annex B countries together reduce emissions by 250 Mt C per year for 70 years (1990–2060), the delay is less than 3 years (70 years $\times$ 0.25 Gt C/6.9 Gt C).

6 The first two IPCC Assessment Reports omitted discussion of adaptation as a result of political pressure from environmental lobbyists; talk about adaptation was construed as acceptance of adaptation as an appropriate policy response (Heal and Kriström, 2002, pp. 35–36; also Kane and Shogren, 2000).

References

Alberta Environmental Protection. (1998). *The Boreal Forest Natural Region of Alberta.* Edmonton, AB: Government of Alberta.

Analysis and Modelling Group (AMG). (2000). *The Economic and Environmental Implications for Canada of the Kyoto Protocol.* Ottawa: National Climate

Change Process. <http://www.nccp.ca/NCCP/national_process/issues/analysis_e.html>.

Darwin, R., M. Tsigas, J. Lewandrowski, and A. Raneses. (1995). *World Agriculture and Climate Change: Economic Adaptations.* AE Report No. 703. Washington, DC: US Department of Agriculture, Economic Research Service.

Ekins, P., and T. Barker. (2001). Carbon Taxes and Carbon Emissions Trading. *Journal of Economic Surveys* 15(3): 325–52.

Fround-Williams, R.J., R. Harrington, T.J. Hocking, H.G. Smith, and T.H. Thomas, eds. (1996). Implications of Global Environmental Change for Crops in Europe. Special issue, *Aspects of Applied Biology* 45.

Heal, G., and B. Kriström. (2002). Uncertainty and Climate Change. *Environmental and Resource Economics* 22(June): 3–39.

Horowitz, J.K. (2001). *The Income-Temperature Relationship in a Cross-section of Countries and Its Implications for Global Warming.* Department of Agricultural and Resource Economics Working Paper No. 01-02. University of Maryland, College Station.

IPCC. (2000). *Land Use, Land-Use Change and Forestry.* Cambridge, UK: Cambridge University Press.

———. (2001). *Climate Change 2001: Mitigation. Technical Summary.* (Contribution of Working Group III to the Third Assessment Report.) Cambridge, UK: Cambridge University Press.

Kane, S., and J. Shogren. (2000). Linking Adaptation and Mitigation in Climate Change Policy. *Climatic Change* 45(1): 75–101.

Kennedy, P. (2002). Optimal Early Action on Greenhouse Gas Emissions. *Canadian Journal of Economics* 35(February): 16–35.

Lise, W., and R.S.J. Tol. (2001). *Impact of Climate Change on Tourist Demand.* FEEM Working Paper No. 48. Amsterdam: Vrije Universitiet.

Maddison, D., and A. Bigano. (2000). *The Amenity Value of the Italian Climate.* Centre for Social and Economic Research on the Global Environment Working Paper. University College London. <http://www.cserge.ucl.ac.uk/publications.html>.

Marland, G., K. Fruit, and R. Sedjo. (2001). Accounting for Sequestered Carbon: The Question of Permanence. *Environmental Science & Policy* 4(6): 259–68.

McKitrick, R., (2001). Mitigation vesus Compensation in Global Warming Policy. *Economics Bulletin* 17(2): 1–6.

McKitrick, R. and R.M. Wigle. (2002). The Kyoto Protocol: Canada's Risky Rush to Judgment. *C.D. Howe Institute Commentary* 169(October). <www.cdhowe.org/pdf/Commentary_169.pdf>.

Mendelsohn, R., W.D. Nordhaus, and D. Shaw. (1994). The Impact of Global Warming on Agriculture: A Ricardian Approach. *American Economic Review* 84: 753–71.

Moore, T.G. (1998). Health and Amenity Effects of Global Warming. *Economic Inquiry* 36(July): 471–88.

Newell, R., and W. Pizer. (2001). *Discounting the Benefits of Climate Change Mitigation: How Much do Uncertain Rates Increase Valuation?* Economics Technical Series. Arlington, VA: Pew Center on Global Climate Change.

Nordhaus, W.D. (1994). *Managing the Global Commons.* Cambridge, MA: MIT Press.

Nordhaus, W.D., and J. Boyer. (2000). *Warming the World: Economic Models of Global Warming*. Cambridge, MA: MIT Press.

Olsen, K.P. Collas, P. Boileau, D. Blain, C. Ha, L. Henderson, C. Liang, S. McKibbon, and L. Moel-a-l'Huissier. (2002). *Canada's Greenhouse Gas Inventory 1990–2000*. Ottawa: Environment Canada, Greenhouse Gas Division.

Pizer, W.A. (1997). *Prices vs. Quantities Revisited: The Case of Climate Change*. Resources for the Future Discussion Paper 98-02. Washington, Resources for the Future.

Stavins, R.N. (2002). Lessons from the American Experience with Market-Based Environmental Policies. In *Harnessing the Hurricane: The Challenge of Market-Based Governance* (forthcoming), ed. J.D. Donahue and J.S. Nye, Jr. New York: Brookings Institution Press.

Suchánek, P., S.L. Shaikh, and G.C. van Kooten. (2002). *Carbon Incentive Mechanisms and Land-Use Implications for Canadian Agriculture*. Working Paper, University of Nevada-Reno.

Takahashi, T., M. Nakamura, G.C. van Kooten, and I. Vertinsky. (2001). Rising to the Kyoto Challenge: Is the Response of Canadian Industry Adequate? *Journal of Environmental Management* 63(2): 149–61.

van Kooten, G.C. (2003). *Climate Change Economics: Prognosis for International Agreement*. Cheltenham, UK: Edward Elgar. 115 pages. <http://web.uvic.ca/~kooten/Downloads/Climate_Primer.pdf>.

van Kooten, G.C. (2000). Economic Dynamics of Tree Planting for Carbon Uptake on Marginal Agricultural Lands. *Canadian Journal of Agricultural Economics* 48(March): 51–65.

van Kooten, G.C., C.S. Binkley, and G. Delcourt. (1995). Effect of Carbon Taxes and Subsidies on Optimal Forest Rotation Age and Supply of Carbon Services. *American Journal of Agricultural Economics* 77(May): 365–774.

van Kooten, G.C., S.L. Shaikh, and P. Suchánek. (2002). Mitigating Climate Change by Planting Trees: The Transaction Costs Trap. *Land Economics* 78(November): 559–72.

Wigle, R. (2001). *Sectoral Impacts of Kyoto Compliance*. Industry Canada Working Paper No. 34. Ottawa: Industry Canada.

8

Regional Adaptation Strategies

Stewart Cohen, Brad Bass, David Etkin, Brenda Jones, Jacinthe Lacroix, Brian Mills, Daniel Scott, and G. Cornelis van Kooten*

DURING 2001, A BARK BEETLE OUTBREAK in British Columbia continued because of extensive fire controls and four years of warm winters, killing lodgepole pine trees in an 8 million hectare area (Reuters, 2001; Morton, 2002). Toronto declared a heat emergency on August 7, while Ontario's daily electricity consumption exceeded 25,000 megawatts for the first time (City of Toronto, 2001; Abbate, 2001). Hunters in Iqaluit noticed that the past few years had been getting warmer and that it takes longer for the ice to form (CBC News, Iqaluit, 2001). Golf courses in southern Ontario were open in December, but ski hills were not (Avery, 2001; Rubin, 2001). An economist with the Canadian Wheat Board estimated a cost of $5 billion for a drought-related 10-million-tonne loss in grain yield (Canadian Press, 2001), while Saskatchewan, Alberta, and Nova Scotia received a total of $3 million in federal drought aid (Brooks, 2001) on top of crop insurance payments. Alberta was considering (again) water transfers from the northern to the southern half of the province, as the government expressed concern about the potential economic impacts of water shortages (Olsen, 2001). In Edmonton alone, the drought of 2001 was reported to have resulted in $4.1 million in damage to trees, shrubs, and sports fields, and an expansion of local watering programs was suggested (Ward, 2002). These are climate impact and adaptation stories from 2001, a warm year among many warm years that have been observed during the last 50 years (see chap. 2). As for

* Opinions expressed in this chapter are those of the authors, and do not necessarily reflect the views of the Government of Canada.

the climate itself, the winter of 2001–02 was the nineteenth consecutive season of above-normal temperatures in Canada (Environment Canada, 2002).

The past decade has witnessed several costly extreme events, including the 1998 ice storm in Ontario and Quebec, the Red River and Saguenay floods, and the 2001 Toronto Heat Emergency (Etkin, 1998; Kerry et al., 1999; IJC, 2000; City of Toronto, 2002). Although recent extreme events cannot be directly attributed to global warming, extreme event frequencies may increase in a future warmer world, leading to more stories like these, and perhaps to more serious challenges as well. Some have argued that warming would be a good thing for a cold country like Canada, but others point to potential ecological and economic disruptions, health impacts, and changing international pressures resulting from the worldwide effects of warming. The Intergovernmental Panel on Climate Change (IPCC) recently concluded that climate change damages would become more pervasive and more costly if warming exceeds a few degrees (IPCC, 2001b). What will happen to Canada and other countries if emission reduction measures are not successful at preventing "dangerous" climate change?

In this chapter, we discuss climate change adaptation in the Canadian context (see chap. 3 for review of global scale concerns and chap. 4 for an overview of projected impacts in Canada). The approach of this chapter is to tell the story of adaptation experiences and future challenges through a series of vignettes. Although many analyses of climate change science and mitigation at national and global levels have been presented in an aggregated way, it is difficult to use that approach when considering climate-related impacts, vulnerabilities, adaptive capacities (economic, institutional, social), and potential options for future response. Adaptation will take place "on the ground" on a regional and local scale. Adapters will be influenced by national and global responses to climate change, but the success of future adaptation measures will depend on regional and local actors, and the perceptions and constraints within which they operate.

What Is Canada's Adaptation Challenge?

Adaptation is defined as an adjustment of a system to an actual or expected stimulus or its impacts. Adaptation is not a new concept (Smit et al., 1999). It is a conscious ongoing process of monitoring, evaluating, and learning to make decisions. These decisions might involve changes to processes, practices, or structures to reduce potential vulnerabilities and damages, or to take advantage of new opportunities that may emerge. Climate adaptation depends on the adaptive capacity of a system, whether it be an ecosystem, human settlement, or a country. A country's adaptive capacity depends on its infrastructure, wealth, institutions, and access to resources, as well as the education and skills of its population. The adaptation challenges faced by any country are also influenced by its vulnerability, the degree to which it is susceptible to, or unable to cope with, adverse effects of climate change and variability (IPCC, 2001a, p. 995).

Another important characteristic of climate adaptation is whether it's autonomous, reactive, or planned. Autonomous adaptation does not constitute a conscious response, but is triggered spontaneously by climate-related stimuli, such as an ecological or market change. Reactive adaptation takes place after impacts have been observed. A planned adaptation is the result of a deliberate policy decision, based on an awareness that conditions have changed or are about to change. In the latter case, we would be talking about anticipatory or proactive adaptation (IPCC, 2001a, p. 982). A summary of adaptation characteristics is shown in Table 8.1. A range of investments is necessary, including intellectual effort, resources, maintenance, and governance. An individual farmer coping with a drought may require different tools and support systems than a province adjusting to changes in forest conditions (fire, pests, etc.), and sometimes climate-related stimuli are superimposed on other challenges, thereby complicating the assessment of the situation.

Table 8.1 Adaptation Characteristics

Criteria	**Characteristics**
Who	Individuals, Communities, Governments, Private Sector
Nature	Autonomous, Planned
Scale	Local, Regional, National, International
Type	Market-based, Behavioural, Legislative, Institutional, Structural, Operational, Technological
Timing	Anticipatory/Proactive, Responsive/Reactive

Source: Authors.

Compared with other countries, Canada should be well positioned to adapt to climate change, given its relatively high levels of wealth and education, and its well-developed infrastructure and resource management systems (IPCC, 2001b). Despite these advantages, however, damage costs from recent climatic events have increased, and conflicts over climate-sensitive resources, such as fish, continue to occur. What are the prospects over the course of the twenty-first century, as populations increase, pressures on land and water resources become more serious, and global economies become more integrated?

As we consider Canada's adaptation prospects, it is important to keep in mind the country's previous experiences with weather and climate variations and trends, as well as current and anticipated trends in economic development. Death rates from extreme weather events and vector-borne diseases have declined during the twentieth century, but weather-related costs have increased dramatically. Insurance costs have also increased, both in Canada and in the United States. In the following sections, we offer several vignettes illustrating Canada's vulnerabilities to current climate, potential implications of scenarios of future climate change, and an assessment of adaptation options.

Canada's Unique Vulnerabilities to Today's Climate

The Canadian climate has always exhibited considerable variability, and Canadians have coped with this. Transportation, housing, recreation, and commercial activities endeavour to align themselves with the opportunities and challenges presented in the climatic conditions found in different regions at various seasons of the year. Adaptation to the current climate, however, comes at a price. Many examples of recent extreme weather events also demonstrate our vulnerabilities. One particular case, the 1998 Ice Storm, is described below, along with a description of how governments and the private sector have used insurance and relief programs in an effort to cope with extreme weather events.

The 1998 Ice Storm: A cautionary tale

In January 1998, an intense ice storm hit eastern Ontario, southern Quebec, and parts of the Maritime Provinces, as well as portions of New York and New England. This was actually a series of three storms that deposited more than 80 millimeters of freezing rain in less than six days. States of emergency were declared in 41 US counties, 3 Canadian cities, and 23 Canadian rural townships and communities (Abley et al., 1998; Harris, 1998). The storm caused power outages that lasted over a week in most areas and much longer in some rural communities. Loss of electric power affected about 4.7 million people. The cities of Montreal and Ottawa were especially hard hit. Indeed, it was fortunate that one grid line remained intact, supplying power to the city of Montreal. Had this failed also (as it easily might have done) the magnitude of the disaster would have increased manyfold.

Did this ice storm occur because of climate change? Increased intensities of winter storms are anticipated for mid-latitude regions as the climate warms (see chaps. 2 and 4), but our purpose is not to suggest that the 1998 storm or other recent weather events have occurred because of human-induced climate change. Such attribution of individual events cannot be determined. What is clear, however, is that there are major forces in society that tend to make us more vulnerable to weather and climate, and that we need to develop better adaptation strategies to deal with them. This case therefore represents an opportunity to learn about climate-related impacts and vulnerabilities in the Canadian context as extreme weather events may occur more often in a warmer future.

How significant were the impacts of the 1998 storm? The total economic impacts have been estimated at C$4.2–C$5.1 billion. This is higher than estimates of insured losses (Table 8.2). Although agriculture and the maple sugar industry were affected, all communities and sectors that relied on electricity suffered. The power loss lay at the heart of the impact. The grid that carried electricity to the city of Montreal suffered massive damage (fig. 8.1). In Quebec, the hardest-hit province, 3,000 km of the distribution system, 16,000 poles, 150 pylons, 4,000 transformers, and 3,000 other structures were damaged. For a province that meets 41% of its total and 68% of its residential energy needs through electricity, this was a catastrophe (Kerry et al., 1999).

Figure 8.1. Damage from the 1998 Ice Storm near Drummondville, Quebec.

Source: Ministry of Environment, Province of Quebec.

There has been some argument about whether the electrical distribution system was or was not designed and maintained to a sufficient level. It certainly lacked resilience. No redundancy or fail-safe designs were incorporated, and when failure occurred, recovery was slow. The weather created additional problems due to an extended freeze period, which made emergency response difficult. This disaster highlighted the vulnerability not only of the energy system but also of our socio-economic system. This was a major

Table 8.2 Estimated Damages from the 1998 Ice Storm

Type of Loss	Canada	United States	Total
Insured losses	Cdn$1.44 billion	US$0.2 billion	US$1.2 billion
Insurance claims	696,590	139,650	835,240
Deaths	28	17	45
Customers without power	4,700,000 (1,673,000)	546,000	5,246,000
Electricity transmission towers/distribution poles toppled	130 / 30,000	?	?
Electric transmission system damage	Cdn$1 billion	?	?
Manufacturing, transportation, communications, and retail business losses	Cdn$1.6 billion	?	?
Forests damaged	?	17.5 million acres	?
Loss of worker income	Cdn$1 billion	?	?
Dairy producers experiencing business disruption	5,500	?	?
Loss of milk	Cdn$7.3 million	US$12.7 million	US$18 million
Agricultural sector (poultry, livestock, maple syrup)	Cdn$25 million	US$10.5 million	US$28 million
Quebec & Ontario Governments	Cdn$1.1 billion		

Source: Adapted from IPCC 2001a.

Note: Based on an analysis conducted by the Canadian Institute for Catastrophic Loss Reduction and the US-based Institute for Business and Home Safety, both insurance industry organizations (Lecomte et al., 1998). Losses as of October 1, 1998 (1 CDN$ = 0.7 US$).

failure of technology. The storm resulted in restrictions to the actual distribution of power. The interruption in the power system disrupted the lives and routines of individuals whose day-to-day functions are based on the use of technology and power.

Over time, we have become increasingly reliant upon technology. This has created a system that is much more efficient under normal circumstances than it was in the past, but suffers to a greater extent when failure occurs. This experience illustrates the need to consider various changes in, for example, design specifications, building codes, and energy distribution systems, in order to meet a changed climate regime. This could actually become part of a revised set of "best practices" with associated economic and legal implications (see chap. 9). Insurance and relief programs will also help meet the adaptation challenges of climate change.

Insurance and relief programs

One of the strategies that society uses to adapt to natural disasters is to share risk. This is most often done through the use of private insurance and government relief programs.

Private insurance allows individuals or groups to transfer risk to other people who buy insurance through the same insurance company that they do, through the payment of annual premiums. The insurance company shares risk by selling insurance policies to many people and by buying insurance from a reinsurance company. Not all risks are readily insurable. The process works best when the risk is spread throughout a large group. Also, the company must be able to estimate the risk so the product can be priced, and there must exist a culture of insurance so that people are motivated to buy it, thereby allowing the company to make a profit. Where risks are strongly differentiated, insurance does not work well. Flood risk is an example of this—people tend to know whether or not they are at risk of flooding (i.e., if they live in a flood plain, they are at risk). If they are not, they do not buy insurance, which tends to make it prohibitively expensive for those at risk. Because of this, residential flood insurance is generally not available in Canada, though it is in the US as part of a government-sponsored program.

The use of insurance by itself does not diminish the cost of a disaster, but insurance does allow better recovery. It is based upon the ethic that the burden of recovery does not lie with society as a whole, but with the individual. In this sense, it is libertarian in nature, as it does not forcibly transfer wealth from others in order to provide relief but respects their property rights. This approach tends to create or amplify social ramps, favouring the wealthy over those who are not able to afford complete insurance. Although sharing risk through insurance is an important tool in the recovery process, it can also have a negative impact in that risky behaviour is encouraged when the cost of that behaviour are viewed as being borne by others. For example, Gilbert White notes that, with respect to the purchase of flood insurance in the United States, "the net effect after several years of practicing such a national policy... may be counter-productive, and the result is an increase in annual losses from floods rather than a decrease. Rather than promoting wise use of floodplains, it might enforce... unwise use" (White, 1999).

Over time, the cost of insurance payouts has increased both within Canada and globally. This is evident in Figures 8.2 and 8.3, which show trends of premium-to-loss ratios for the insurance industry. Disaster relief payments from the federal government to the provinces were actually declining during 1970–1994, but rose sharply during 1995–1999 (fig. 8.4). In addition, recent government crop insurance payments have varied between $50 million in 1996 and $1 billion in each of 1988 and 1989, which were years of major drought. Payments for droughts in 2001 and 2002 were $570 million and one billion, respectively (Bonsal et al., 2003). The drought of 2003 may result in payments of similar magnitude.

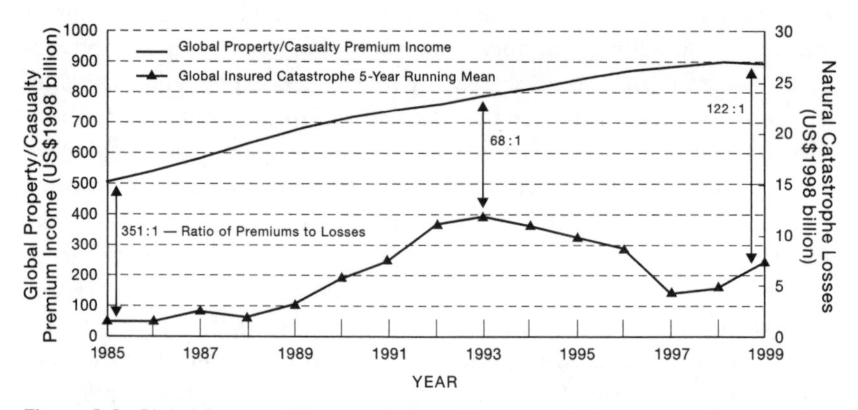

Figure 8.2. Global Insured Weather-Related Catastrophe Losses vs. Property/Casualty Premium Income

Source: IPCC, 2001a.

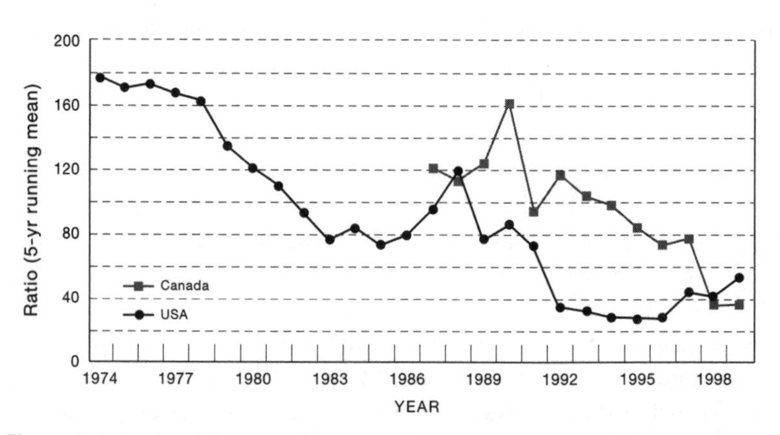

Figure 8.3. Ratio of Property/Casualty Premiums to Insured Weather-Related Catastrophe Losses in Canada and the US

Source: IPCC, 2001a.

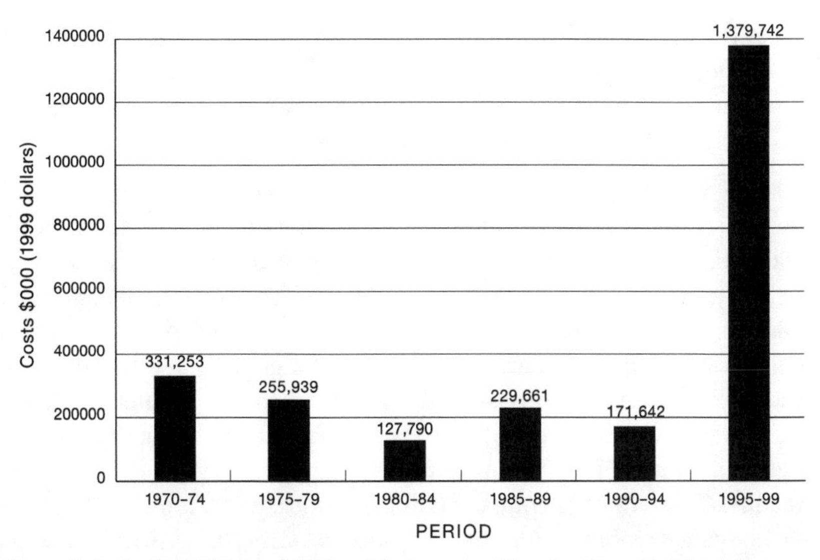

Figure 8.4. Audited Totals of Claims Made under Disaster Financial Assistance Arrangements (DFAA)

Source: Data from Office of Critical Infrastructure Protection and Emergency Preparedness.

Note: These figures do not include costs of provincial assistance to sub-DFAA threshold events (i.e., less than $1 per capita provincial population).

Both public and private sectors are now considering how better to mitigate their losses in the future, by creating more disaster-aware and resilient communities. In Canada, the main process by which the insurance industry works towards this goal is through research and lobbying efforts by the Institute for Catastrophic Loss Reduction (www.iclr.org), which held a series of workshops across Canada in 1998 in order to develop a national mitigation strategy (ICLR, 1998). The following principles underlie their approach to mitigation:

- The threat of severe weather is increasing; nevertheless, sustained action can reduce catastrophic losses. Hazard assessment and risk identification are cornerstones of catastrophic loss mitigation.
- Solid, applied research provides an essential foundation for effective action to reduce future losses.
- Those who knowingly choose to assume greater risk must accept an increased degree of responsibility for their choice.
- Communication with the public before a peril strikes is an important means of reducing losses.

- Local and individual actions are the most effective means of reducing the loss of life and property.
- Partnership is the best approach to resolving shared problems, particularly public safety concerns.

Government relief programs are based upon an ethic that emphasizes the greater social good. The idea is that the relief provided to victims provides a benefit to the larger society (this is particularly true if an industry such as agriculture has been affected, by drought, for example) and that there is a moral obligation to assist fellow citizens in need. A counter-argument based upon egalitarian principles would suggest the redistribution of resources resulting from disaster relief programs to those in greatest need, such as the homeless, while one based upon libertarian assumptions would argue that such a redistribution of resources violates the rights of the wealthy. Clearly though, one cannot ignore disaster victims in Canadian society, which embraces the ethic that we should help those who have suffered through little or no fault of their own. These programs are often essential if communities are to recover from rare, unpredictable, disastrous events. If overused or used unwisely, however, such programs can create cultures of dependency, where people expect to be looked after and do not accept individual responsibility. Overuse of government relief programs can then increase social vulnerability at the local level, even while these programs act to increase resilience in the larger society.

The Office of Critical Infrastructure Protection and Emergency Preparedness (OCIPEP) coordinates federal relief programs (www.ocipep.gc .ca/) under the Disaster Financial Assistance Arrangement (DFAA), which allows provinces to make a claim for federal financial resources to assist with reconstruction following disasters. Claims can be made when provincial expenditures exceed $1 per capita of population. Ottawa then pays 50% of the next $2 per capita, 75% of the following $2 beyond that, and finally 90% of any remainder. Joint crop insurance programs also exist to assist the agriculture industry. OCIPEP is partnering with the ICLR in continuing the process of developing a national mitigation strategy. The latest consultation began on 26 June 2001 with the purpose of developing "a national strategy through which all levels of government and stakeholders can co-operate effectively to evaluate, prioritize and implement risk and impact reduction measures aimed at reducing the risk and impact of disasters" (Eggleton, 2001).

Canada has a culture that strongly emphasizes insurance and government relief as tools to share risk and assist recovery from natural disasters.

Increased costs during the mid-1990s, however, have led government and industry to reconsider the effectiveness of these programs and to look for other tools that might be used to reduce vulnerability.

Dialogues on Adaptation

When assessing the potential implications of future climate change, researchers face the uncertainties of projecting how temperature, precipitation, and other climate parameters would change for various regions in Canada. The climate change crystal ball may not be completely opaque at the global scale, but climate modellers have clearly stated that there are concerns related to downscaling from the global scale to regional scales (see chap. 2). However, "what if" case studies using science-based scenarios of climate change still provide the opportunity to learn about impacts and adaptation. It is important that these cases consider scenarios of regional development that could alter land and water use, thereby affecting vulnerability to climate-related risks and regional abilities to adapt.

Throughout North America, many stories are emerging on the challenges of adaptation to climate change. Some of these were illustrated in the IPCC Third Assessment Report (fig. 8.5). Some of these cases, such as the 1998 ice storm in Ontario and Quebec, illustrate experiences from recent extreme weather events. Others, such as Great Lakes water levels (see chap. 4) or Okanagan/Columbia water management, reflect outcomes of "what if" scenario exercises, using model-based climate change scenarios as a starting point, but eventually reaching into concerns about regional vulnerabilities and development choices in the face of changing national and global circumstances. How should communities, businesses, and governments respond? Who chooses adaptation options for regions? Will adaptation exist only in a reactive mode, or will it be possible and feasible to adapt in an anticipatory manner, despite uncertainties about future climatic and socio-economic conditions? Canadians will have to cope with a diversity of situations as they consider their options in the face of climate change.

Canada's National Park System

National Parks are a celebrated part of the landscape and an important part of Canada's national identity. Analysis of climate change impact studies on Canadian ecosystems indicates that climate change will have important implications for ecological integrity in Canada's national parks and will pose a significant policy and management challenge to Parks Canada.

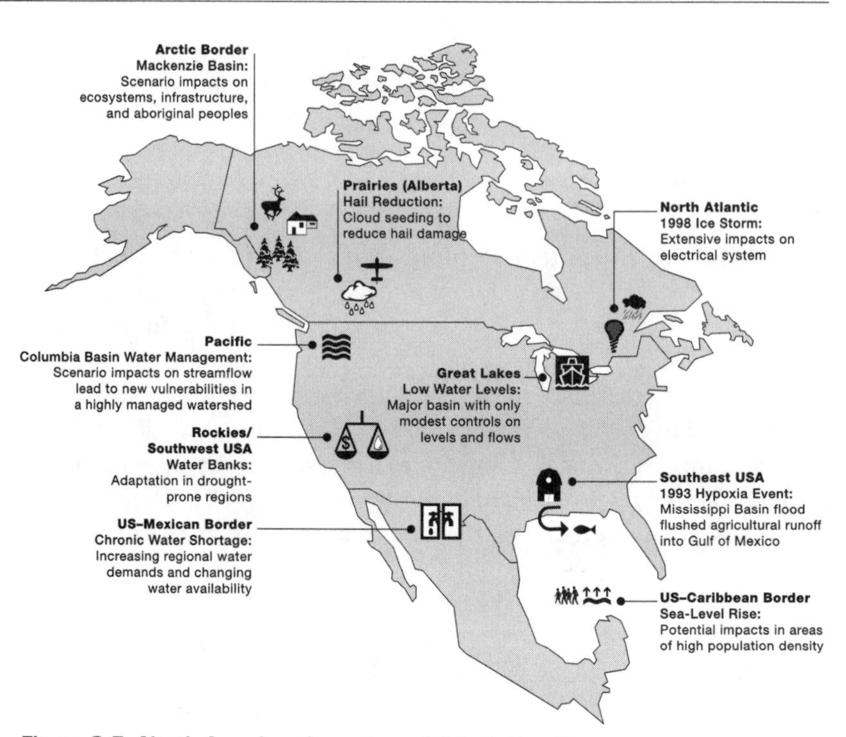

Figure 8.5. North American Impacts and Adaptation Cases
Source: IPCC, 2001a.

Canada's national park system is predicated on natural regions defined in part by vegetation classification. Using results from equilibrium global vegetation models driven with multiple climate change scenarios, Scott et al., (2002) examined potential changes in biome representation in Canada's national parks. Regardless of the vegetation-modelling scenario used, substantial changes in biome representation occurred in the national park system. In five of six vegetation scenarios, a novel biome type appeared in over half of the national parks (e.g., grasslands developing in a park dominated by boreal forest). The proportional representation of biomes in the national park system also changed, with diminishing representation of northern biomes (tundra, tundra/taiga, and boreal forest) and additional representation of more southerly biomes (temperate evergreen and temperate mixed forests in particular). Although equilibrium vegetation modeling results can only indicate the magnitude and probable trajectory of vegetation change, and cannot predict the eventual future distribution and composition of biomes in Canada, these and other vegetation modelling results (Neilson, 1998; Cramer et al., 2001) demonstrate that a reassessment of the system plan is necessary to consider contingencies for climate change.

Policy and planning sensitivities also exist at the individual park level. For example, the stated purpose of Riding Mountain National Park (RMNP) is to "Protect for all time the ecological integrity of a natural area ... representative of the boreal plains and mid-boreal uplands." As Figure 8.6 illustrates, the park's mandate would be untenable in the long term as vegetation modelling projects the eventual loss of boreal forest in the park.

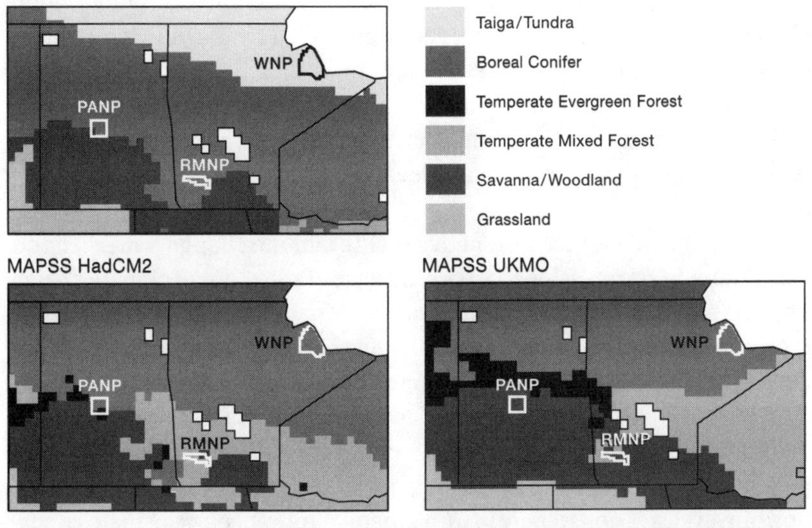

Figure 8.6. Implications of Potential Biome Shifts in Canada's National Parks
Riding Mountain National Park (RMNP), Prince Albert National Park (PANP), and Wapusk National Park (WNP). MAPSS global vegetation model (Neilson, 1995) run with current climate conditions and forced with Had CM_2-ghg and UKMO doubled-CO_2 climate change scenarios (see Scott et al., 2002 for details).
Source: Scott et al., 2002.

Superimposed on the projected biome distribution change would be a range of other regional climate change impacts that will also affect the ecological integrity of the national parks (see Scott and Suffling, 2000, for a summary). For example, of the nine national parks with coastal areas on the Atlantic Ocean, seven were rated as highly sensitive or moderately sensitive to physical changes resulting from projected sea level rise, including coastal erosion and salinity changes that could possibly degrade some key marine, dune, tidal pool, salt marsh, and estuary habitats. In Canada's Arctic parks, projected changes in growing season, permafrost, insect regimes, and sea-ice would have significant ecological impacts. For example, Wapusk National Park (WNP) was established in 1996 largely to protect one of the world's largest denning and gathering places for polar bears. Polar bears in western Hudson Bay are already being threatened by sea-ice reductions (Stirling et al., 1999), and sea-ice models project conditions in Hudson Bay that would limit the bears' ability to hunt to the extent that they would be extirpated from the very park established for their protection.

Scott et al., (2002) and Suffling and Scott (2002) have identified policy and planning sensitivities that raise fundamental questions about the long-term role of the national parks and Parks Canada in an era of climate change. The growing body of evidence compiled by Hughes (2000), IPCC (2001a, 2001b), and Walther et al., (2002), indicating that some species have already begun to respond to climate changes during the twentieth century, demonstrates that climate change is not just a conservation problem in the distant future but has relevance to current conservation policy and planning.

Parks Canada is being proactive on the climate change issue. In 2000, Parks Canada commissioned a scoping level assessment of climate change impacts and has used the report as professional learning document. Since 1997, the number of national parks that list climate change as a significant ecological stressor in their official management plans has increased from 4 to 29 (out of 39). Parks Canada has also initiated research to examine adaptation options and how climate change could be integrated into the park establishment and park management processes. Parks Canada cannot, however, act unilaterally to implement comprehensive climate change contingency plans, as this would require legislative changes to the National Parks Act that must be initiated by Canada's Parliament.

Impacts on Streamflow in the Okanagan/Columbia

Water resources, their management, and their use are known to be sensitive to variations in climate (Hofmann et al., 1997; IPCC, 2001a). Potential implications of climatic change for the Okanagan region of British Columbia include earlier onset of spring peak flows, by as much as 4–6 weeks (fig. 8.7). The projected peak is often lower than current peak flows, especially at low elevation watersheds. All areas in the region show loss of snowpack. Winter flow increases, while summer flow decreases. There is no consensus on changes to total annual flow (Hamilton, 2001).

These results are consistent with those of Hamlet and Lettenmaier (1999) and Miles et al., (2000) for the Columbia Basin as a whole. The Columbia Basin case suggests a tendency towards both earlier onset of

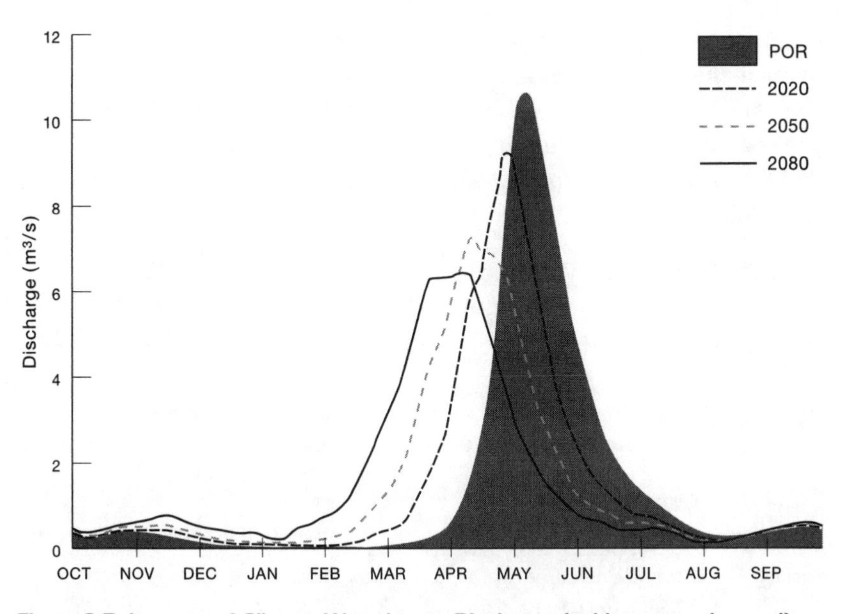

Figure 8.7. Impacts of Climate Warming on Discharge (cubic metres/second) at Dave's Creek, Okanagan Region, British Columbia

Based on IS92a emissions scenario and resulting climate change projection from the Canadian climate model CGCM1. POR = Period of record. Emissions scenarios are described in chapter 2.

Source: Hamilton, 2001.

peak flows and reduced annual flows, which would affect the reliability of the water management system to meet various goals, including provision of irrigation services, firm energy production, flood control, and "biological flow" to support fish habitat. Despite a high level of development and management in the basin, vulnerabilities would still exist and impacts could still occur (Cohen et al., 2000).

A dialogue on adaptation challenges and opportunities in the Columbia Basin has recently been initiated by the University of Washington (Miles et al., 2000) and the Sustainable Fisheries Foundation (SFF, 2002). A case study on adaptation in the Okanagan region is also focusing on dialogue with regional water interests as a way of bringing researchers and stakeholders together to share knowledge and generate new ideas about ways to adapt to the uncertainty of climate change. In the Okanagan, the prospects of an earlier spring snowmelt accompanied by a longer growing season and lower minimum streamflow present a challenge. Preliminary discussions with Okanagan stakeholders have revealed a wide array of views about how the region might adapt to climate change, with a preference for structural measures, particularly intervention to prevent impacts. Other popular alternatives include developing alternative uses for resources and changing land-use plans (Table 8.3). Regarding implications of adaptation choices, considerable attention was given to water licensing, flow regulation through dams, and potential restrictions on development (Cohen and Kulkarni, 2001).

Table 8.3 Stakeholder Views on Adaptation Options in the Okanagan Region, British Columbia

Cluster	Type of Response	Examples
Structural	Intervention (24)	Snowmaking, upland dams
	Defensive expenditure (10)	Fire protection, buy out water licenses
	Priced regulation (6)	Water pricing, metering
	Regulate access (4)	Conservation
	Regulate land/water use (17)	Control irrigation, protect watersheds
Social	Alternate resource use (16)	Small-scale water storage, grey water
	Educate to change (4)	Risk reduction
	Lifestyle change (8)	Water consumption for grass
	Land-use planning (12)	Urban densification
Other	Hands off (5)	Fewer dams
	More research (5)	Flow regimes
	Improve governance (3)	Agency integration

Source: Cohen and Kulkarni, 2001.

Note: Numbers in parentheses indicate number of times an option was identified in focus group sessions.

Some options appear to be more popular than others, but this dialogue is just beginning. How would information on costs of adaptation measures influence regional preferences? What about the side effects on fisheries or low income groups? The preliminary discussions described above were a short-term exercise in the use of scientific models to form the foundation for a regional dialogue on climate change impacts and adaptation. For this effort to be effective over the long term, there needs to be a sustained investment in this process.

Toronto Hot Weather Response Plan

Statistical studies have demonstrated strong positive relationships between indicators of heat stress and mortality (Smoyer et al., 2000a, 2000b; Kalkstein, 1993; Kalkstein and Davis, 1989; Chestnut et al., 1998). The occurrence of extreme heat waves and reporting of associated heat-related deaths and hospital admissions, most notably in Chicago in 1995 (Whitman et al., 1997; Semenza et al., 1999), have encouraged public health agencies in several large southern Ontario municipalities to consider establishing and, in at least two cases (City of Toronto and Regional Municipality of Waterloo), to implement response plans to mitigate the effects of heat on health. Research on the potential impacts of anthropogenic climate change has also heightened awareness of heat-related health risks and generally suggests that heat-stress conditions may become more frequent and severe in the future in mid-latitude cities like Toronto (Chiotti, 2002; IPCC, 2001a).

The City of Toronto developed and implemented a hot weather response plan in 1999 whereby one-day forecasts for humidex values of 40°C or more provided by Environment Canada trigger the mobilization of special city programs and services (City of Toronto, 2002). Examples of such measures provided by either the City of Toronto or funded community organizations include

- provision of alert notices and available services to the media and to organizations and groups that care for the most vulnerable (i.e., Community Care Access Centres, Toronto Public Health high-risk clients, hostels, seniors' agencies);
- opening of air-conditioned emergency cooling centres during a heat emergency;
- delivery of bottled water to libraries, community centres, and other public facilities;
- provision of educational material to the general public and to seniors and other vulnerable groups (i.e., through Red Cross heat information line, community libraries, major pharmacy chains, Ontario Community Support Association);
- extension of Anishnawbe Health Toronto Street Patrol operating hours to communicate risks to homeless people and provide public transit passes for travel to cooling centres;
- provision of emergency home medical and environmental assessments and response to people at risk of developing heat-related illnesses (Toronto Emergency Medical Services and Red Cross); and
- activation of emergency operations centres if a prolonged heat emergency occurs (City of Toronto, 2002).

Difficulties associated with accurately predicting peak humidex values in part prompted the City of Toronto to support the development of a Toronto heat-health alert system (Kalkstein, 1999), which was used to activate the plan commencing in 2001. The system is based upon a spatial synoptic classification method (Kalkstein et al., 1998) that associates air mass types with elevated mortality (relative to normal or average daily mortality for that location). Heat *alerts* are issued by the Medical Officer of Health when an oppressive air mass is forecast and the probability of excessive mortality is greater than 65%; the Mayor declares a heat *emergency* when a 90% or greater probability of excess deaths is predicted (City of Toronto, 2002).

The new, synoptic-based alert system was piloted during the summer of 2001. Four alerts were issued, covering a total of six days in June, July, and August, and a heat emergency was declared for three additional days in August (City of Toronto, 2002). The alert/emergency days are shown in Figure 8.8 together with Environment Canada humidex advisory days and temperature data for a downtown Toronto observing station. Between May 1 and September 30, daily maximum temperatures reached or exceeded 30°C and corresponding minimum temperatures failed to drop below 20°C on 15 days. From August 6–9 the daily maximum temperature exceeded 35°C.

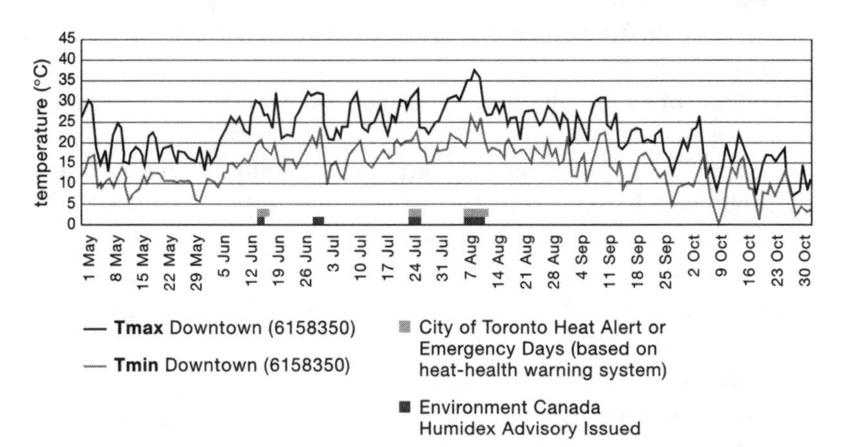

Figure 8.8. City of Toronto Heat Alert/Emergency Days, Environment Canada Humidex Advisory Days, and Temperature Data for a Downtown Toronto Observing Station (MSC-6158350) May–October 2001

Source: Authors.

Media reports identified heat as a contributing factor in five deaths in or near Toronto (Jones, 2001a, 2001b). Over 500 calls to the Red Cross heat information hotline were received during the alert and emergency days—21 of these led to patients being transported to hospital emergency departments by Toronto Emergency Medical Services (City of Toronto, 2002).

Over 1,000 people took advantage of the four cooling centres that were opened during the August 7–9 heat emergency (City of Toronto, 2002).

The heat-health warning system was deemed a success given that the actual death toll appeared to fall below the predicted excess four to five deaths per day during the heat emergency (Jones 2001b). Media reports suggested that the lack of increased demands for hospital emergency room services during the June and July alerts and again during the final two days of the August 7–9 heat emergency—relative to the first day—also indicated the program's success and people's willingness to adopt protective measures (Jones, 2001a, 2001b). While the anecdotal reports in the media are positive and useful feedback, it is important that rigorous evaluations be completed using hospital records and other data over several years to demonstrate the utility of this adaptive strategy.

Green Roof

Urban areas are sensitive to extremes of precipitation and extreme heat. Extreme precipitation events can result in flooding but more commonly in a strategy called combined sewer overflow (e.g., using the sewage system to handle the overflow runoff from storms, flushing pollutants from sewage into a receiving water body). Extreme heat in the summer can increase the rate of smog formation, triggering respiratory problems, heat stress, and increased consumption of electricity for air conditioning, which, if generated by coal or oil, will increase fossil-fuel emissions and air pollution. The sensitivities result from the replacement of vegetation with surfaces that are impervious to moisture and absorb more solar energy than plants, thus re-radiating more energy as heat.

Figure 8.9. Green Roof, Toronto City Hall

Source: photo by The Cardinal Group, Inc.

Green roof infrastructure provides an opportunity to replace part of the vegetation surface (fig. 8.9). Using green roof infrastructure to reduce the urban heat island is important for several reasons. Rooftops provide additional space for vegetation to complement urban forestry, and in high-density commercial areas roofs may provide far more surface than is available for trees at ground level. Roof surfaces are also quite warm in the summer; temperatures of 60°C are not unusual. Green roof infrastructure can reduce stormwater runoff as the tops of buildings can retain anywhere from 40–100% of rainwater or snowmelt depending on the leaf surface area, the soil depth, the amount of rainfall, and the duration of the event.

Green roof infrastructure is also a technology; it is more than just putting soil on a roof and planting seeds, a practice that would most likely damage the roof and result in water leaking into the building. The most common technology uses a water- and root-repellent membrane, a drainage layer for excess moisture, and, if necessary, lightweight soil alternatives to reduce the weight of the roof and avoid costly retrofits. The vegetation canopy can vary from an alpine meadow to a food crop, from a typical English garden to other types of ecosystems, depending on the weight of the roof, the budget, and the needs of the building owner.

The green roof can provide many other benefits, such as doubling the life of the roof, providing wildlife habitat to preserve biodiversity, creating urban green space, offering new economic opportunities, reducing energy consumption, and increasing property values. Interestingly enough, these benefits have been known for thousands of years, and some variation of green roof infrastructure has been in existence in Europe since the Roman Empire. New York City once had a vibrant green roofscape to provide green space to the new class of urban apartment dwellers.

The green roof industry has "grown" significantly in Europe, particularly in Germany, where governments have since 1989 put in place legislation that requires new buildings to minimize certain environmental impacts. At a stakeholder workshop held in Toronto in 1998, several barriers to widespread adoption in Canada were identified. These included a lack of government support, a lack of technical information, a lack of awareness, and cost. To address the information barrier, a research program has been launched with Environment Canada, the National Research Council, the City of Toronto, the roofing industry, and other industrial stakeholders. The program has obtained funding from a variety of sources, including the Climate Change Action Fund, to construct two research sites and to develop simulation models for evaluating the reduction of stormwater runoff, energy consumption, and the urban heat island.

A green roof can be combined with other types of roofs or roof-top features, such as solar panels, rainwater collection devices, and skylights. Solar roofs and green roofs make an interesting contrast. Both types use incoming solar energy; the former converts it to heat or electricity, and the latter prevents solar energy from being converted to waste heat, and thus cools the atmosphere above the roof and the floor beneath it as well. The degree to which these two technologies could be combined on a roof to maximize benefits is an optimization and an architectural problem. For example, what is the trade-off between stormwater retention and solar electricity? Perhaps this combination will spur the wider adoption of both of these technologies as cities cope with the environmental impacts of climate change.

Prince Edward Island: Coastal impacts and the Confederation Bridge

An example of anticipatory adaptation is the dialogue on the design of the Confederation Bridge, the link between Prince Edward Island and New Brunswick. It was anticipated that the bridge would have a lifespan of 100 years, and so could be affected by climate change. When the bridge design was being considered, concerns were raised regarding the implication of sea-level rise (Environment Canada, Atlantic Region, 1990). This led the bridge designers to consider a one-metre sea-level rise (Jacques Whitford Environment Limited, 1993), as well as potential stresses from wind, waves, and ice. The minimum height of the approaches to land was to be high enough that sea ice would not be trapped by the bridge spans and sea spray would not affect the bridge surface.

The coastal zone impacts of sea-level rise combined with storm surges may become a problem for Charlottetown. Storm surge events in January and October 2000 caused damage to wharves, roads, bridges, business, and homes. The January storm caused nearly $1 million in insured losses, but this likely underestimates the true cost. In a scenario of the January 2000 storm surge plus a 0.7 metre sea-level rise, public infrastructure and private property of over $200 million would be at risk of flooding (Environment Canada et al., 2001). Several adaptation options have been identified, mostly focused on retreat from vulnerable shorelines, including legislated setback distances, building code changes, and conservation of natural ecosystems. Some areas may require protection where justified in terms of the ratio of cost to value protected. It was recognized, however, that there might not be a single best option.

Adaptation to climate change by prairie agricultural producers

In a series of studies conducted during the 1980s, Dr. Louise Arthur and her colleagues at the University of Manitoba determined that Prairie agriculture could benefit from global warming (Arthur, 1988; Arthur and Abizadeh,

1988; Mooney et al., 1991; Mooney and Arthur, 1990). While areas of southern Alberta and southwest Saskatchewan would experience moisture deficits that would make anything but livestock grazing infeasible in the absence of irrigation, areas where moisture would not be a limiting factor could benefit because of longer projected growing seasons. Models project that warming will shift the US Corn Belt into the Canadian prairie region, with farmers in Manitoba and parts of Alberta and Saskatchewan potentially benefiting from opportunities to grow more highly valued crops. In the northern grain belt, where soil conditions permit, it may be feasible to grow grains in place of trees. While agriculture in the prairie region may experience an overall gain from climate change, serious dislocations may also occur as some farmers can be expected to experience losses. Further, even farmers that stand to gain by growing a different mix of crops are likely to incur greater financial risks in order to take advantage of any such opportunities. Are they prepared to adapt to climate change, and, if so, what strategies might they adopt?

In order to answer questions related to the likely strategies agricultural producers might adopt, a survey was sent to a random sample of western Canadian farmers in July 2000 (Suchánek, 2001; van Kooten et al., 2002). The average farm in the sample has 1,869 acres, with a minimum of 160 acres and a maximum of 10,000 acres; 90% are under 4,000 acres. Farms had been under the respondents' ownership for an average of 33 years, and 70% of farms had been inherited, indicating long-term familial commitment. This is likely to continue as 57% of the respondents indicate an intention to pass their farms on to their descendants. Most are full-time farmers, with only 38% indicating that off-farm employment supplemented the household's income. Almost all respondents were familiar with the subject of climate change, and the majority of respondents (84%) knew about possibilities for mitigating climate change through activities such as conservation tillage, reduced tillage summer fallow, and planting trees.

Farmers were presented with a permanent climate change scenario: a 3°C increase in average daily growing-season temperatures, 10% lower available soil moisture, a growing season that is 15% longer, and milder winters with similar amount and variability in snowfall. Only 10% stated that these conditions would cause them to leave agriculture altogether; nearly 60% indicated that they would continue in the business. However, nearly one-third of respondents were uncertain about whether or not they would remain in agriculture. Given a range of options, farmers chose first to seek advice, either about new crops or about how to manage weather-related risks, or both. Planting of genetically modified crops that are most drought and pest resistant (a contentious strategy given environmental and food-safety concerns), and greater reliance on a mix of crops that includes livestock, also featured prominently as adaptation strategies. Interestingly, greater use of chemical fallow was more likely to be adopted than increased tillage fallow as an adaptation strategy for conserving soil moisture. Farmers indicated they would be unlikely to rely more on irrigation, likely because irrigation is not generally practiced in Canada's grain belt (with the exception of southern Alberta). Somewhat surprisingly, farmers indicate that they would not increase plantings of crops grown in the US Corn Belt, despite its projected shift towards the north and west into Canada (fig. 8.10)

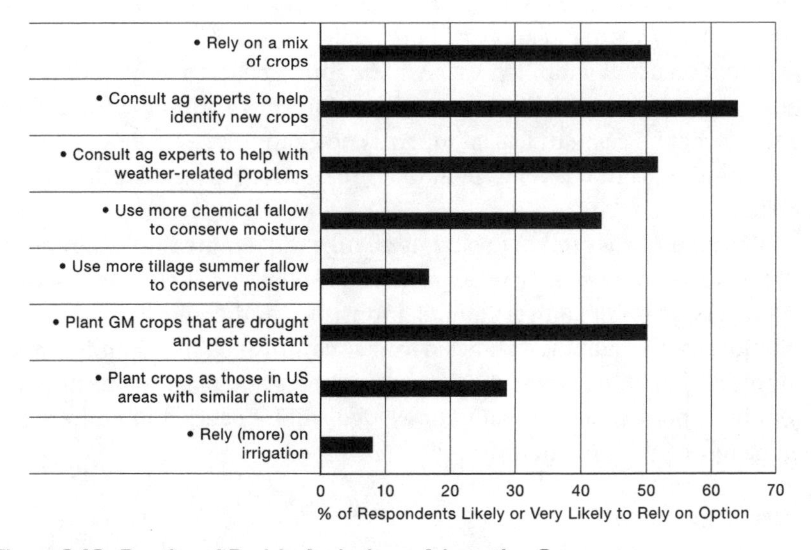

Figure 8.10. Results of Prairie Agriculture Adaptation Survey
Source: Suchánek, 2001; van Kooten et al., 2002.

Can Adaptation Play on Canada's Climate Change Team?

The ultimate objective of any response to climate change is the avoidance of "dangerous anthropogenic interference" with the atmosphere. Reduction of greenhouse gas emissions has attracted much of the attention of decision makers and the public, and there are a number of opportunities and challenges associated with successfully reducing emissions. Any definition of "dangerous" is a value judgment that will vary according to the context of individual places, societies, and economies. Through the vignettes presented in this chapter, we have tried to show that vulnerabilities and adaptive capacities are unique to each regional circumstance. We suggest here that adaptation measures ought to be part of the climate change portfolio of responses, in that such measures could reduce vulnerabilities and increase capacities to adapt. This would, in effect, buy some time for emission reduction mechanisms to be effectively implemented.

Climate-related adaptation has occurred in the past, and will happen in the future as the climate changes. In order to support an adaptation program, we need a mechanism for monitoring the impacts of climatic events and attempts to adapt to them. Studies of recent extreme events, such as the 1998 Ice Storm, have suggested how development choices have contributed to the creation of new vulnerabilities on the ground. There are also many examples, such as the Winnipeg floodway (IJC, 2000) and the Toronto Hot Weather Response Plan that illustrate how planning can lead to reduction of vulnerability. A systematic approach to documenting adaptation successes and failures, in conjunction with additional future-oriented "what if" case studies of impacts and adaptation, would be a useful contribution to the development of a broad strategy for adapting to climate change.

Successful adaptation to climate change will require the collaborative effort of researchers and a wide range of stakeholders, including resource managers, planners, governments, industries, and communities. Scientific, local, and traditional knowledge have contributed to our understanding of the dynamic relationship between climate and place. A holistic view that incorporates all forms of knowledge will be needed to address the challenge of adapting to future climate change.

Acknowledgments

The authors would like to thank Quentin Chiotti, Roger Street, Tina Neale, Ian Burton, Grace Koshida, and Linda Mortsch for their comments and suggestions; Keith Keddy and Gary Lines for providing documents on the Confederation Bridge and impacts on Prince Edward Island; and AIRG's university partners at University of British Columbia, University of Waterloo, University of Toronto, York University, and Université de Québec à Montréal for their ongoing support.

References

Abbate, G. (2001). Relief from heat expected in Maritimes and Prairies. *Globe and Mail,* 8 August 2001.

Abley, M., A. Arpin, D. Johnston, I. Peritz, and J. Mennie. (1998). The Review: The Ice Storm of '98. *The Montreal Gazette,* 24 January 1998, 1–6.

Arthur, L.M. (1988). The Implications of Climate Change for Agriculture in the Prairie Provinces. *Climate Change Digest* 88: 1–13.

Arthur, L.M., and F. Abizadeh. (1988). Potential Effects of Climate Change on Agriculture in the Prairie Region of Canada. *Western Journal of Agricultural Economics* 13: 215–24.

Avery, R. (2001). Praying for a White Christmas: Warm Weather Has Ski Resort Operators Worried. *Toronto Star,* 4 December 2001.

Bonsal, B.A., A. Chipanshi, C. Grant, G. Koshida, S. Kulshreshtha, E. Wheaton, and V. Wittrock. (2003). Canadian Droughts of 2001 and 2002: Climatology, Impacts and Adaptations: Summary Report. E. Wheaton, V. Wittrock, and S. Kulshreshtha, eds. Saskatoon, SK: Saskatchewan Research Council (SRC), SRC Publication No. 11602-33E03.

Brooks, P. (2001). Drought Aid Welcomed; Ottawa Pledges $300 Million to Ease Farmers' Water Woes. *The Sunday Herald,* 9 December 2001.

Canadian Press. (2001). Farmers Struggle to Survive as Drought Dries Up Hopes. *Winnipeg Free Press,* 15 October 2001.

CBC News, Iqaluit. (2001). Weather Conditions Worry Nunavut People. P. Bell radio interviews in Igloolik and Iqaluit. Transcript. *World Report,* 24 December 2001.

Chestnut, L.G., W.S. Breffle, J.B. Smith, and L.S. Kalkstein. (1998). Analysis of Differences in Hot-Weather-Related Mortality across 44 US Metropolitan Areas. *Environmental Science and Policy* 1: 59–70.

Chiotti, Q. (2002). *Towards an Adaptation Action Plan: Climate Change and Health in the Toronto–Niagara Region.* Report submitted to the Adaptation Liaison Office, Natural Resources Canada. Toronto: Pollution Probe.

City of Toronto. (2001). Mayor Declares Heat Emergency for Toronto. 7 August 2001.

City of Toronto (2002). Update on Hot Weather Response Plan. Staff report dated 21 March 2002. <http://www.city.toronto.on.ca/legdocs/2002/agendas/committees/hl/hl020408/it003.pdf>.

Cohen, S. and T. Kulkarni, eds. (2001). *Water Management and Climate Change in the Okanagan Basin.* Project A206, submitted to the Adaptation Liaison Office, Climate Change Action Fund, Natural Resources Canada. Vancouver: Environment Canada/University of British Columbia.

Cohen, S.J., K.A. Miller, A.F. Hamlet, and W. Avis. (2000). Climate change and resource Management in the Columbia River Basin. *Water International* 25(2): 253–72.

Cramer, W., A. Bondeau, F.I. Woodward, I.C. Prentice, R. Bettes, V. Brovkin, P. Cox, et al., (2001). Global Response of Terrestrial Ecosystem Structure and Function to CO_2 and Climate Change: Results from Six Dynamic Global Vegetation Models. *Global Change Biology* 7: 357–73.

Eggleton, A. (2001). Department of National Defence Press Release. 26 June 2001. <http://www.ocipep.gc.ca/>.

Environment Canada. (2002). Climate Trends and Variations Bulletin for Canada Winter 2001/02 Temperature and Precipitation in Historical Perspective. <http://www.msc.ec.gc.ca/ccrm/bulletin/>.

Environment Canada, Natural Resources Canada, Department of Fisheries and Oceans, and Dalhousie University Centre of Geographic Sciences. (2001). Coastal impacts of climate change and sea-level rise on Prince Edward Island. CCAF A041. Dartmouth, NS: Climate Change Action Fund.

Environment Canada, Atlantic Region. (1990). *Environment Canada's Position on the Northumberland Strait Crossing Project.* Dartmouth: Environment Canada.

Etkin, D. (1998). Climate Change and Extreme Events: Canada. In *Canada Country Study: Climate Impacts and Adaptation,* vol. 8, *National Cross-Cutting Issues,* ed. N. Mayer, and W. Avis, 31–80. Ottawa: Environment Canada.

Hajat, S., R.S. Kovats, R.W. Atkinson, and A. Haines. (2002). Impact of Hot Temperatures on Death in London: A Time Series Approach. *Journal of Epidemiology and Community Health* 56: 367–72.

Hamilton, S. (2001). Hydrology. In Cohen and Kulkarni, 26–29, 48–68.

Hamlet, A.F., and D.P. Lettenmaier. (1999). Effects of Climate Change on Hydrology and Water Resources in the Columbia River Basin. *Journal of the American Water Resources Association* 35(6): 1597–624.

Harris, E. (1998). Struck Powerless. *Canadian Geographic* 118: 38–45.

Hofmann, N., L. Mortsch, S. Donner, K. Duncan, R. Kreutwizer, A. Kulshreshtha, A. Piggott, S. Schellenberg, B. Schertzer, and M. Slivitzky. (1998). Climate Change and Variability: Impacts on Canadian Water. In *Canada Country Study: Climate Impacts and Adaptation,* vol. 8, *National Sectoral Issues,* ed. G. Koshida and W. Avis, 1–120.

Hughes, L. (2000). Biological Consequences of Global Warming: Is the Signal Already Apparent? *Trends in Ecology and Evolution* 15: 56–61.

Institute for Catastrophic Loss Reduction (ICLR). (1998). *Better Protecting Canadians from Natural Hazards.* Toronto: Institute for Catastrophic Loss Reduction.

Intergovernmental Panel on Climate Change (IPCC). (2001a). *Climate Change 2001: Impacts, Adaptation and Vulnerability.* Contribution of Working Group II to the Third Assessment Report of the Intergovernmental Panel on Climate Change, ed. J.J. McCarthy, O.F. Canziani, N.A. Leary, D.J. Dokken, and K.S. White. Cambridge, UK: Cambridge University Press.

Intergovernmental Panel on Climate Change (IPCC). (2001b). *Climate Change 2001: Impacts, Adaptations, and Vulnerability: Technical Summary.* IPCC Working Group II Third Assessment Report. Cambridge, UK: Cambridge University Press.

International Joint Commission (IJC). (2000). *Living with the Red: A Report to the Governments of Canada and the United States on Reducing Flood Impacts in the Red River Basin.* Ottawa and Washington: International Joint Commission.

Jacques Whitford Environment Limited. (1993). Environmental evaluation of SCI's Proposed Northumberland Strait Crossing Project. Dartmouth: Environment Canada.

Jones, V.C. (2001a). Record Heat Takes Its Toll. *Toronto Star,* 9 August 2001, p. A1.

———. (2001b). Two Seniors Die of Heatstroke. *Toronto Star,* 11 August 2001, p. A1.

Kalkstein, L.S. (1999). Heat-Health Watch Systems for Cities. World Meteorological Organization Bulletin. *WMO* 48(1): 69.

———. (1993) Health and Climate Change: Direct Impacts in Cities. *Lancet* 342: 1397–99.

Kalkstein, L.S., and R.E. Davis. (1989). Weather and Human Mortality: An Evaluation of Demographic and Inter-Regional Responses in the U.S. *Annals of the Association of American Geographers* 79: 4–64.

Kalkstein, L.S., and J.S. Green. (1997). An Evaluation of Climate/Mortality Relationships in Large US Cities and the Possible Impacts of a Climate Change. *Environmental Health Perspectives* 105: 84–93.

Kalkstein, L.S., S.C. Sheridan, and D.Y. Graybeal. (1998). A Determination of Character and Frequency Changes in Air Masses Using a Spatial Synoptic Classification. *International Journal of Climatology* 18: 1223–36.

Kerry, M., G. Kelk, D. Etkin, I. Burton, and S. Kalhok, (1999). Glazed Over: Canada Copes with the Ice Storm of 1998. *Environment* 41(1): 6–11, 28–33.

Lecomte, E., A.W. Pang, and J.W. Russell. (1998). *Ice Storm '98*. Ottawa: Institute for Catastrophic Loss Reduction; Boston: Institute for Business and Home Safety.

Miles, E.L., A.F. Hamlet, A.K. Snover, B. Callahan, and D. Fluharty. (2000). Pacific Northwest Regional Assessment: The Impacts of Climate Variability and Climate Change on the Water Resources of the Columbia River Basin. *Journal of the American Water Resources Association* 36(2): 399–420.

Mooney S., and L.M. Arthur. (1990). Impacts of 2 × CO_2 on Manitoba Agriculture. *Canadian Journal of Agricultural Economics* 38: 685–94.

Mooney, S., S. Jeffrey, and L.M. Arthur. (1991). *The Economic Impacts of Climate Change on Agriculture in the Canadian Prairie Provinces.* Winnipeg: Department of Agricultural Economics, University of Manitoba.

Morton, B. (2002). It's Not Cold Enough to Stop Pine Beetle: Northern BC Forests Need −40 for a Week, Industry Spokesman Says. *Vancouver Sun,* 29 January 2002.

Neilson, R. (1995). A Model for Predicting Continental-Scale Vegetation Distribution and Water Balance. *Ecological Applications* 5: 362–85.

———. (1998). Simulated Changes in Vegetation Distribution Under Global Warming. In *The Regional Impacts of Climate Change: An Assessment of Vulnerability.* A Special Report of IPCC Working Group II, ed. R. Watson, M. Zinyowera, and R. Moss, 441–56. Cambridge, UK: Cambridge University Press.

Olsen, T. (2001). Water Shortage Alarms Province: Diversion Plan Possible. *Calgary Herald,* 27 December 2001.

Pal, K. (2002). Assessing Community Vulnerability to Flood Hazard in Ontario: A Case Study Approach. *Canadian Journal of Water Resources,* submitted.

Rainham, D.G.C. (2000). *Atmospheric Risk Factors of Human Mortality.* MSc Thesis, University of Alberta.

Reuters. (2001). Western Canadian Beetle Epidemic Spreading Rapidly. 30 November 2001.

Rubin, J. (2001). Toronto's Blooming Warm; Gardens, Golfers Spring to Life as Record High Nears. *Toronto Star,* 5 December 2001.

Scott, D., J. Malcolm, and C. Lemieux. (2002). Climate Change and Biome Representation in Canada's National Park System: Implications for System Planning and Park Mandates. *Global Ecology and Biogeography* 11: 475–84.

Scott, D., and R. Suffling. (2000). *Climate Change and Canada's National Parks.* Toronto: Environment Canada.

Semenza, J., J. McCullough, D. Flanders, M. McGeehin, and J. Lumpkin. (1999). Excess Hospital Admissions During the 1995 Heat Wave in Chicago. *American Journal of Preventive Medicine* 16: 269–77.

Smit, B., I. Burton, R.J.T. Klein, and R. Street. (1999). The Science of Adaptation: A Framework for Assessment. *Mitigation and Adaptation Strategies for Global Change* 4: 199–213.

Smoyer, K., L. Kalkstein, J. Scott Greene, and Y. Hengchun. (2000b). The Impacts of Weather and Pollution on Human Mortality in Birmingham, Alabama, and Philadelphia, Pennsylvania. *International Journal of Climatology* 20: 881–97.

Smoyer, K., D. Rainham, and J. Hewko. (2000a). Heat Stress Mortality in Five Cities in Southern Ontario: 1980–1996. *International Journal of Biometeorology* 44: 190–97.

Stirling, I., N.J. Lunn, and J. Iacozza, (1999). Long-Term Trends in the Population Ecology of Polar Bears in Western Hudson Bay in Relation to Climate Change. *Arctic* 52: 294–306.

Suchánek, P. (2001). Farmers' Willingness to Plant Trees in the Canadian Grain Belt to Mitigate Climate Change. Unpublished MSc thesis: Faculty of Agricultural Sciences, University of British Columbia.

Sustainable Fisheries Foundation (SFF). (2002). Towards Ecosystem-Based Management: Breaking Down the Barriers in the Columbia River Basin and Beyond. Proceedings of Conference held in Spokane, 27 April–1 May 2002. <http://www.sff.bc.ca/2002.html>.

Suffling, R., and D. Scott. (2002). Assessment of Climate Change Effects on Canada's National Park System. *Environmental Monitoring and Assessment* 74: 117–139.

van Kooten, G.C., S. Shaikh, and P. Suchánek. (2002). Mitigating Climate Change by Planting Trees: The Transaction Costs Trap. *Land Economics* 78: 559–72.

Walther, G., E. Post, P. Covery, A. Menzel, C. Parmesan, T. Beebee, J. Fromentin, O. Hoegh-Guldber, and F. Bairlein. (2002). Ecological Responses to Recent Climate Change. *Nature* 416: 389–95.

Ward, J. (2002). Dry Skies Cost Us Millions in 2001. *Edmonton Sun*, 12 January 2002.

White, G. (1999). A Conversation with Gilbert F. White. *Environmental Hazards* 1: 53–56.

Whitman, S., G. Good, E.R. Donoghue, N. Benbow, W. Shou, and S. Mou. (1997). Mortality in Chicago attributed to the July 1995 heat wave. *American Journal of Public Health* 87(9): 1515–18.

Legal Contraints and Opportunities: Climate Change and the Law

Alastair R. Lucas

Introduction

Climate change law now consists almost exclusively of international law. To this point in Canada, little has been done at the national level either to implement international law obligations or to create additional legal rights and obligations to address climate change. Of course, for a federal country like Canada *national* must be understood as federal or provincial according to the constitutional distribution of powers to legislate.

This chapter looks at climate change law by sketching the overall legal framework and identifying some more specific legal tools as well as potential legal constraints. At present, climate change law, particularly at the national level, is largely speculative. There is a range of possible policies, approaches, and instruments that can be used in relation to climate change. Policy and instrument choices will depend on federal and provincial processes that are now under way and that will ultimately be governed by political, economic, and technical factors. The law is relevant to this process primarily for the constraints that it may impose, including international law constraints, constitutional constraints, constraints in existing legislation that may point to necessary new legislation, and perhaps liability constraints. International law cannot be changed by Canadian governments, nor, apart from the difficult and unlikely possibility of amendment under the process prescribed by the constitution, can constitutional requirements be altered. Legislative constraints can be removed by amendment, but various policy, intergovernmental, and political factors may present barriers. Once made, climate change policy choices may then require legislation for implementation. Only then will climate change law have taken root in Canada. Thus, apart from fundamental legal norms that are constitutionalized or quasi-constitutionalized as part of international agreements, the law does not lead. It is essentially responsive so that legislation may be viewed as the formal authoritative expression of government policy.

It is important, particularly for understanding the Kyoto Protocol debate, to clarify the nature of international climate change law and how this international law relates to Canadian national law and the Canadian legal system. The first part of the chapter will do this by focusing on the primary international law instruments, the UN Framework Convention on Climate Change (UNFCCC) and the Kyoto Protocol. The implications of another specific area of international law—international trade law—will also be assessed.

The national legal framework will then be discussed, beginning with relevant constitutional law constraints. More specific elements of the legal framework at the federal and provincial level will be assessed to identify statutory provisions, primarily of a regulatory nature, that could be used as a basis for climate change initiatives and decisions. Some are specific statutory powers that may be used directly; most are enabling provisions that permit federal or provincial cabinets or ministers to make regulations or other subordinate legislation for climate change purposes. Finally, potential liability issues concerning climate change actions will be discussed.

International Law and Canadian Law

Canada's signing of the Kyoto Protocol in December 1997 did not make it a law binding on Canadians; nor do the Protocol's obligations become enforceable legal requirements for Canadian individuals and corporations through its ratification by the federal cabinet. Nonetheless, the Protocol does not lack legal force. Its foundation instrument, the UNFCCC, was ratified by Canada in 1994. The Kyoto Protocol is therefore binding as it has been ratified (Kyoto Protocol, 1997, Art. 24.1); but, like the UNFCCC, it binds only Canada the nation state as party to the Protocol recognized by international law.[1] However, the binding force in international law may itself be uncertain since it depends largely on the non-compliance provisions contained in the international agreement itself, and the Protocol merely authorizes the Conference of the Parties (COP) to "approve effective procedures and mechanisms" and adopt these by an amendment to the Protocol (Kyoto Protocol, 1997, Art. 18). It is a matter of negotiation by the COP, which has agreed[2] on a Compliance Committee system for implementation and determination of non-compliance, with ultimate penalties of submission and completion of a compliance plan, and suspension of emissions trading and international project emission credit rights. Ultimately, the legal force of the UNFCCC obligations and those of the Protocol is a matter of international relations and the expectations and pressures of the international community.

Should the Kyoto Protocol come into force, making its obligations binding on Canadians would require action by Canadian lawmakers. It would be the responsibility of Canada to enact the national legislation necessary to implement the Protocol.[3]

The UNFCCC and the Kyoto Protocol as International Legal Instruments

The UNFCCC and the Kyoto Protocol are international multilateral agreements. Canadian climate change commitments arise from Canada's having become a party to these international legal instruments in the manner outlined above. Canadian commitment began with signing, along with 153 other countries, the UNFCCC at the Earth Summit in Rio de Janeiro in June 1992. This commitment was confirmed by the federal cabinet's ratification of the UNFCCC.[4] The act of ratification created Canada's international legal obligation to comply with the terms of this international agreement.

The central commitment of parties to the UNFCCC is to adopt national policies and take corresponding measures with the "aim" (UNFCCC, 1994, Art. 4.2 [a] [b]) of reducing greenhouse gas emissions to 1990 levels. Though developed countries were required to "take the lead" (UNFCCC, 1994, Art. 3.1, 4.2 [a]) by estimating emissions and reporting these through emissions inventories that form part of "national communications," no specific emission limitations for parties to the Convention were established. What was established, and this is characteristic of complex modern international conventions, is a process and a set of institutions to carry out the refinement and implementation of the Convention. These institutions include a Secretariat to manage the ongoing operations under the Convention (UNFCCC, 1994, Art. 8), particularly facilitation of the main decision-making body under the Convention, the Conference of the Parties (UNFCCC, 1994, Art. 7), and expert review teams to assess national communications and to coordinate and support the COP.

Conferences of the Parties, beginning with the April 1995 Berlin Conference (COP 1), began to work toward strengthening the general greenhouse gas (GHG) emission reduction commitments by developing quantified limits and targets and by establishing specific time frames.[5] The Kyoto Protocol was a product of the third Conference of the Parties (COP 3), after the Parties had agreed[6] to work toward a protocol or other legal instrument to address emissions in the post-2000 period. The Kyoto Protocol is thus a legal instrument that is subsidiary or adjunct to the UNFCCC. For Canada, it requires a 6% emission reduction over 1990 levels by the 2008–2012 period specified in the Protocol. However, to make it operative as a matter of international law, a process similar to that used for the UNFCCC is required, namely ratification as specified in the Protocol by at least 55 countries representing 55% of 1990 GHG emissions.[7] In line with the UNFCCC, the Kyoto Protocol establishes the Convention of the Parties to the UNFCCC as its decision-making institution (Kyoto Protocol, 1997, Art. 1.1). Consequently, following the 1997 Kyoto Conference, a series of Conferences of the Parties[8] have worked toward the establishment of operating rules and guidelines, particularly for carbon accounting, measurement and monitoring, and implementation of the Kyoto mechanisms of joint implementation, the Clean Development Mechanism, and emissions trading (Grimeuld, 2001; Mimms, 2001).

International Trade Agreements

Canadian obligations under other international treaties may potentially conflict with national GHG reduction actions taken in the context of the UNFCCC and the Kyoto Protocol. This is particularly the case for obligations under Canada's international trade agreements—The World Trade Organization (WTO, 1994) and the North American Free Trade Agreement (NAFTA, 1992). These multilateral agreements impose fundamental trade obligations, including national treatment (imported products are treated no differently than domestic products, and quantitative restrictions are prohibited), most favoured nation (no discrimination among products of different countries), and progressive tariff reduction. Countries are also required to maintain transparency by providing reasonable access to trade laws and regulations and trade actions taken. There are, of course, exceptions, including protection of national health and the environment, provided there is non-discrimination and provided such actions are not merely disguised trade restrictions.

In principle, actions taken under an environmental agreement such as the Kyoto Protocol could, if inconsistent with a prior trade agreement, be challenged by countries who are parties to the trade agreement but not the environmental agreement. However, in the case of the UNFCCC and Kyoto Protocol, it is arguable that countries that ratified the 1992 UNFCCC, to which the Kyoto Protocol is related, may be considered to have waived their rights under the 1994 WTO in relation to inconsistencies (Vienna Convention on the Law of Treaties, 1969). Alternatively, it is possible that rights under the second international agreement prevail in the event of conflict. For Canada, this view would support prevalence of the UNFCCC in relation to the US because the UNFCCC came into force after the 1988 Canada-US Free Trade Agreement. GHG emission reduction actions would be covered by the UNFCCC because the Protocol merely quantifies and provides mechanisms for realizing the fundamental GHG reduction commitments under the UNFCCC.

In any event, analysis to date suggests that it is unlikely that properly structured GHG reduction actions under the Kyoto Protocol can be characterized as trade measures (Stewart et al., 1996). Though it is arguable that emission reduction units are "goods" or "products" under the WTO, it is more likely that Kyoto allowances are transactable units created by governments to facilitate compliance with an international agreement, rather than goods in trade. However, as a cautionary matter it has been suggested (Petsonk, 1999) that countries should avoid discrimination against other countries or regions in defining and structuring the emission reduction units created under the different Kyoto mechanisms, and in particular should avoid quantitative restrictions.

It is possible that allocation of emission allowances, which may differ from country to country and, within Canada, between provinces and classes of industries, would constitute an impermissible subsidy under the trade agreements. Allowances, once created, will acquire value, and surplus units could be sold so that allocation, particularly initial *gratis* allocation or partly *gratis* allocation, may be regarded as a subsidy (Dunham, 2002). However, remedy would be indirect, via countervailing duties. Perhaps more significantly, to characterize allocation of allowances as a subsidy would raise broader issues about national environmental regulatory regimes that trade dispute panels may wish to avoid. Otherwise, the effect could be that sovereign decisions by Canadian federal and provincial governments to regulate certain contaminant sources but not others, or to impose less onerous legal requirements on some contaminant emitters than on others, may be characterized as creating actionable subsidies.

The National Legal Framework

There is no legislation in force at either the federal or provincial level that directly addresses climate change or the UNFCCC and Kyoto Protocol.[9] It is likely that legislation will be required to implement the Kyoto Protocol

GHG reduction commitments. However, the nature and form of this legislation is unclear. It depends on precisely what measures are adopted by the federal government and by provinces. Action to date has shown that much can be done without legislation through policy and fiscal actions and through public education and voluntary initiatives. Ultimately, legislation will be required to specify GHG caps, to allocate emission rights, to establish emission trading rules and enforcement mechanisms, and to link international reduction measures and trading to the national system.

Intergovernmental Agreements

Another legal instrument, agreements among governments and between governments and private industry, will also be necessary. Intergovernmental agreements will establish the allocation of Canada's total GHG reduction obligations among provinces and the federal government. This will not be an easy negotiation. It is a matter of intergovernmental politics, but an important card in the negotiation will be held by the level of government considered most likely to have the constitutional jurisdiction to legislate on this matter. In particular, is there federal jurisdiction to legislate an allocation and establish a national framework for an emissions trading system? This brings us to that ultimately eye-rolling Canadian exercise of attempting to assess constitutional competence.

Constitutional Competence

The legal framework for emission reduction to meet Kyoto Protocol commitments has both federal and provincial components. There is no power under the Constitution Act, 1867, to permit federal pre-emption of legislative authority in relation to national environmental protection (Hogg, 1997); nor, despite the federal power to negotiate and ratify international treaties such as the UNFCCC and the Kyoto Protocol, is there federal constitutional power to fully implement these treaties.[10]

The provinces have exclusive powers to legislate in relation to "property and civil rights in the province," (Constitution Act, 1867, s. 92[13]), "local works and undertakings," (s. 92[10]), and "management and sale of public lands in the province" (s. 92[5]). With several important exceptions, relevant federal heads of power tend to be narrower and more specific. These include "navigation and shipping" (s. 91[10]), interprovincial and international transportation facilities (s. 92[10][a]), Indian lands (s. 91[24]), and various other categories of federal lands, including the northern territories and the offshore marine belts (s. 109 and Third Schedule). This means that emission-producing activities within provincial boundaries are, with some exceptions such as interprovincial transportation, within the exclusive legislative jurisdiction of the provinces.

However, there are also several more sweeping federal powers, namely those exercised over interprovincial, national and international trade and commerce (Constitution Act, 1867, s. 91[2]), and the criminal law (s. 91[27]). Finally, there is the general federal power to make laws for the "peace, order and good government" of Canada in relation to matters of "national concern" (s. 91[29],and concluding clause of s. 91). The latter has been severely restricted by judicial interpretation, though the decisions have recognized exclusive federal power to regulate the quality of marine waters[11] and atomic energy,[12] which suggests a basis for relevant federal environmental jurisdiction. However, more recently, in *R. v. Hydro Quebec*[13] the Supreme Court of Canada largely dismissed the federal power of peace, order, and good government as a basis for federal environmental jurisdiction, but upheld the toxics regulation provisions of the federal Canadian Environmental Protection Act under the criminal law power. The court emphasized that the environment as such is not a sufficiently distinct subject of legislation for constitutional jurisdiction purposes. This suggests that in assessing the constitutional validity of federal environmental legislation, the court must begin with the catalogue of federal legislative powers under section 91 of the Constitution Act, 1867, rather than focusing on national concern under the power for peace, order, and good government.

What this means for constitutional jurisdiction over greenhouse gas reduction measures is far from clear. With certain exceptions,[14] it is futile to speculate about legislative jurisdiction over subjects, policies, or even particular kinds of statutory schemes or instruments. The only relevant questions for judicial review concern the constitutional validity of specific legislative provisions. Everything may depend on how statutory schemes are constructed and the particular language used, and until the Supreme Court has decided, constitutionality remains at least somewhat speculative. Positional politics apart, this is very likely a significant factor in Alberta's decision to pull back from an announced intention to mount a constitutional challenge to federal Kyoto Protocol implementation jurisdiction[15] (Ferguson, 2002).

A major exception referred to above is federal power over direct and indirect taxation (Constitution Act, 1867, s. 92[13]). Thus, carbon tax measures, unless overlaid with a significant regulatory apparatus, should present no constitutional difficulty for the federal government. However, federal marketable emission rights schemes are more problematic. Emission rights, whether considered to be a form of property or rights arising from "civil rights" to contract, appear to fall within the provincial "property and civil rights" jurisdiction. It is likely that trading schemes are too complex to be addressed by the federal government under its criminal law power even as elaborated by the Supreme Court in *R. v. Hydro Quebec.* There is a possibility that emissions trading or greenhouse gas control could meet the peace, order, and good government criteria. It may be a matter of "national concern," particularly if the "provincial inability"[16] of foot-dragging provinces prevents concerted action through provincial measures. But the "scale"[17] of impact of such a federal scheme on core provincial powers over property, natural resources, and local industry weighs against federal jurisdiction. Federal trade and commerce jurisdiction is also unlikely since trade is an instrument rather than a fundamental aspect of an emissions trading scheme. Furthermore, as indicated above, the precise scheme and its statutory language must be the basis for this constitutional analysis.

Not surprisingly, opinions of scholars are divided (Castrilli, 1999; Rolfe, 1998; Bachelder, 2000; Bankes, 1991), but none have concluded that the federal government has constitutional jurisdiction broad enough to permit an optimal scheme to be tailored. Consequently, there is at least a likelihood that the federal government lacks constitutional authority to legislate national standards and the necessary framework for a national emissions trading system. The result is that federal-provincial agreement is necessary, and constitutional jurisdiction is not a strong card for either negotiating side.

Municipal Jurisdiction

Constitutionally, Canadian municipalities are not a distinct jurisdictional level. Their legislative and corporate powers are derived from provincial municipal government statutes and until recently their bylaw powers were thought to be based on strictly interpreted powers listed in their statutes. However, in 2000 the Supreme Court of Canada made it clear that broadly worded "peace, order, good government, health and general welfare" powers conferred by municipal government statutes must be interpreted purposively and given a "benevolent construction that recognizes the responsibility of elected municipal councils and permits them to perform their legitimate role as community representatives."[18] In the *Spraytech* case, the court confirmed the validity of a pesticide regulation bylaw passed by the Town of Hudson under this "general welfare" power. This development broadens the jurisdiction of municipalities to enact bylaws designed to encourage or require local action on climate change.

Legislation

Though the Canadian environmental legal framework does not address greenhouse gas emissions or global warming explicitly, certain direct regulatory statutory provisions can be identified that are potentially relevant to greenhouse gas reduction. There are also statutory enabling provisions providing authority that may be used to make regulations or other subordinate legislation to implement greenhouse gas reduction objectives and approaches such as emissions trading. It is apparent that most of the relevant statutory provisions, whether federal or provincial, are in this latter category. In addition, there is potential for voluntary measures, which many Canadian governments apparently consider not to require any legal framework at all (Lucas, 2000). This issue is discussed below.

Direct Regulation

One example of direct regulation is the Energy Efficiency Regulations (Canada Gazette Part II, 1995) made under the federal Energy Efficiency Act (Canada, 1992). These establish standards for a wide range of electrical products listed in the Schedule to the Act, and thus contribute to more efficient use of electricity and stabilization of emissions in electricity generation. A second example is the Ozone Depleting Substances Regulations (Canada Gazette Part II, 1995) under the Canadian Environmental Protection Act (CEPA) (Canada, 1999, Part 5). These regulate chlorofluorocarbon manufacture and sale in a manner that implements Canada's obligations under the Montreal Protocol on Substances that Deplete the Ozone Layer (Montreal Protocol, 1987). They are made under the broader CEPA toxic substances control scheme that is discussed below. Coordinated provincial regulations prohibit manufacture and sale in provinces of ozone depleting substances.

Some direct regulation is actually implemented through voluntary means. Under CEPA, the federal Cabinet has power to make regulations concerning concentrations of constituents and physical properties of fuels, fuel characteristics, and transfer and handling (Canada, 1999, s. 140). A separate Act, the Motor Vehicle Fuel Consumption Standards Act (Canada, 1981) was enacted in 1981. However, it has not been proclaimed in force and is said to exist as "an alternative to voluntary fuel efficiency standards"

(Transportation Issue Table, 1998). Proclamation has been unnecessary because the Canadian motor vehicle industry has undertaken to meet standards equivalent to those in the US (Transport Canada, 1996).

Enabling Provisions

An "enabling" provision is a statutory power that authorizes executive decision-makers, usually the federal cabinet, to make specific regulations concerning those matters specified in the enabling provision. The regulations noted above concerning ozone depleting substances are an example of this subordinate legislation created through the use of enabling powers. Where enabling powers exist, matters relevant to global warming can be regulated quickly and directly by cabinet or ministerial regulation without the need to follow the more transparent, and therefore, lengthier legislative process.

Provincial Permit Systems

Permit-type regulatory systems that target the release of contaminants into the natural environment, and that are backed by quasi criminal offence provisions, are central to the Canadian environmental regulatory framework. Such schemes exist in all of the provinces and at the federal level under the Canadian Environmental Protection Act. Constitutional conflict has been largely defused by CEPA's toxics regulation implementation process, which has limited its scope[19] (Canada, 1999, Part 5), and by its mechanism for withdrawal in the face of equivalent provincial provisions (Canada, 1999, s. 10). Although provincial schemes either potentially encompass GHGS or contain enabling provisions to include them, the principal GHGS have not been regulated under the permit schemes.

CEPA Regulation of Toxics

Similarly, at the federal level under CEPA, although the mechanism and enabling powers exist to regulate major GHGS, only chlorofluorocarbons have been the subject of specific regulations. CEPA's process requires identification and "toxicity" assessment according to the broad definition of "toxic"[20] in the Act, scheduling of substances found to be toxic, and promulgation of regulations to regulate and control scheduled substances. It is a process that could potentially be used to establish standards in the form of ambient caps or emission limits and to regulate the principal GHGS.

However, the CEPA provisions illustrate that these enabling powers may be subject to conditions, including threshold criteria and process[21] requirements.

International Air Pollution

CEPA also contains enabling provisions that authorize the federal Cabinet to make regulations with respect to emission sources in order to control or prevent international air pollution (Canada, 1999, ss. 166–174). This includes the regulation of contaminant sources likely to result in the violation of an international agreement entered into by Canada in relation to control and abatement of pollution. Potentially, this would include GHG sources based on Canada's UNFCCC commitments and now ratified on the Kyoto Protocol.

Emissions Trading

Both provincial and federal statutes contain enabling provisions that authorize executive regulations to facilitate emissions trading. An Alberta provision concerning economic instruments merely empowers the minister to make regulations and establish programs and other measures for the use of economic instruments and market-based approaches, "including... emission trading" (Alberta, 1992, s. 13). More detailed sections of the CEPA recognize specific aspects and issues of emissions trading schemes (Canada, 1999, ss. 322, 326). The CEPA specifies "tradeable units" and conditions for their "creation, distribution exchange, sale, use, variation or cancellation," as well as a public registry for the system, conditions for use of and participation in the system, and record-keeping and reporting. It also contemplates identification of substances or products released, description and nature of tradeable units, baselines and maximum emissions limits, and testing and monitoring. There are overriding environmental and health protection ministerial powers to order conditions, to suspend or cancel trading, and to invalidate any trade, where ministers are of the opinion that the trade or use of a tradeable unit:

a) has or may have an immediate or long-term harmful effect on the environment;
b) constitutes or may constitute a danger to the environment on which human life depends; or
c) constitutes or may constitute a danger in Canada to human life or health. (Canada, 1999, s. 327).

Notwithstanding these discretionary ministerial powers in CEPA, these enabling powers and the more open-ended enabling provisions in most provinces do provide sufficient legal authority for the establishment of emissions trading schemes. Most of these provisions create flexibility to accommodate simple cap-and-trade systems or more complex trading systems with, for example, several caps to recognize carbon offsets.

Environmental Impact Assessment

None of the federal or provincial environmental impact assessment statutes mentions greenhouse gas emissions. However, environmental assessment processes that provide for comprehensive environmental assessment and review, including public review of projects likely to cause significant adverse environmental effects, can accommodate advance assessment of potential GHG emissions. The definitions of "environment" and "environmental effect" include climate changes likely to have adverse effects on the natural environment (Alberta, 1992, s. 1). To this point however, GHG emission information filed by project proponents has not been quantitative, and assessment has accordingly been limited to evaluating proponent commitments to, for example, offset GHG emissions and prepare voluntary GHG reduction plans.[22]

Incentives for Energy Alternatives

Most alternative energy research and development subsidy programs are spending initiatives based on general departmental enabling powers that do not require specific legislation. Federal initiatives include Natural Resources Canada energy technology research programs such as the Canada Centre for Mineral and Energy Technology (CANMET), Energy Technology Centre (CETL), the CANMET Energy Diversification Research Laboratories, the Program of Energy Research and Development (PERD), and the Renewable Energy Deployment Initiative, as well as the Technology Early Action Measures (TEAM) Program that is a component of the federal Climate Change Action Fund initiative. Support for private sector action includes seed funding for Ballard Power's innovative hydrogen cells, support for ethanol research, and a variety of demonstration projects.

Alternative energy development support that is tied into energy regulatory systems is the subject of special legislation. An example is Alberta's Small Power Research and Development Act (Alberta, 1988), which provides grid access and price incentives for up to 125 megawatts of electricity produced by small power projects including wind, hydro, and biomass generating facilities. Programs approved by regulators in British Columbia and Ontario provide a system where electricity generated by small hydro facilities is purchased at avoided cost.

Capacity for Legal Adaptation

It is clear that while some elements of the legal framework necessary to facilitate policy changes and to provide legal tools for effective Kyoto Protocol implementation are in place, much remains to be done. It is equally clear that there are constraints on institutional ability to make the necessary changes. In theory, legislatures are sovereign, at least within their constitutional legislative competence: but it is this matter of constitutional competence that presents the first problem.

Federal-Provincial Arrangements

As discussed above, it is unlikely that the federal government has constitutional competence sufficient to enact legislation to address fully GHG reduction, including Kyoto Protocol implementation. Consequently, some form of federal-provincial agreement or understanding is necessary. Although intergovernmental discussions were ongoing at a technical and policy level, the question of an appropriate timetable for ratification of the Kyoto Protocol triggered hostile political exchanges in 2002. Following high profile expressions of concern by a group of provincial premiers during a Team Canada European trade mission (Bueckert, 2002), initial federal intention to ratify in 2002 gave way to a more consultative approach. However, the provinces are by no means in accord (Bueckert, 2002) and other interests, including municipalities (Adam, 2002), have expressed a variety of positions. Alberta threatened, then backed away from, a legal challenge to federal constitutional jurisdiction to implement the Protocol (Ferguson, 2002).

What must be settled is the allocation of the total GHG reduction requirement to the respective provinces, as well as related roles, legislative and regulatory action, and financial obligations. The difficulty in assessing economic costs associated with the necessary GHG reduction is a major problem. Provincial economic situations differ significantly, with western energy-producer provinces in a particularly vulnerable position. Alberta's high calculation of costs to its economy has been criticized by federal officials as outdated and failing to account for use of the Kyoto flexibility mechanisms and for climate change costs such as damage caused by extreme weather (Jaimet, 2002). Studies produced by NGOs have concluded that a net benefit will result from Kyoto implementation (Tellus Institute, 2002). All of this signals the beginning of serious government-to-government negotiations.

Legislative Modification

To a significant extent, enabling legislation already exists that could be used to implement federal and provincial GHG obligations. Subsidy program authorization is likely to be found in federal and provincial departmental statutes that empower energy and environment ministers or related regulatory agencies. Similarly, tax measures such as deductions and accelerated depreciation can be accommodated under existing income tax legislation. As discussed above, enabling provisions permit the establishment of emission trading schemes. All of these enabling provisions will, however, have to be assessed when specific proposals emerge. Some existing powers may be too narrow and will require legislative amendment. Other powers may be lacking completely, and new legislation will be necessary to provide enabling powers or to authorize directly measures such as emissions trading. Careful review of legislation will be required in each province and at the federal level to ensure that GHG reduction measures have the necessary jurisdictional basis. This will avoid painful surprises if statutory powers or decisions under them are challenged by GHG emitters or by third parties such as affected citizens, corporations, and NGOs, through judicial review.

Existing environmental approval regulatory systems will have to be assessed to determine, and if necessary clarify, their relationship with GHG emission reduction schemes. For example, to what extent should GHG emissions and participation in GHG schemes, including international actions, be assessed in approval applications for GHG-producing facilities such as oil-sands plants? Do provincial regulatory agencies such as the Alberta Energy and Utilities Board have the legal authority to consider and base approval terms and conditions on GHG offsets purchased from other provinces or from sources outside Canada?[23]

Similarly, it is possible that recognition of international GHG credits is, as a matter of purposive statutory interpretation,[24] outside the scope of emissions trading and other GHG measures that are based on enabling powers in existing environmental statutes. Recognizing GHG credits would then necessitate legislative amendments or new legislation. However, the Supreme Court of Canada has confirmed in the *Spraytech* case[25] that international law, including international agreements such as the UNFCCC, and customary international law, including the "precautionary principle," must be taken into account in purposive statutory interpretation.

Voluntary Initiatives

Canadian governments have attached great significance to voluntary compliance approaches in planning for and taking action to reduce GHG emissions. In its 1995 National Action Program on Climate Change, the Government of Canada included "implementation of voluntary actions" as a major element of its four strategic directions in meeting its commitments under the UNFCCC to reduce GHG emissions (Environment Canada, 1995). The cooperative industry-government Voluntary Challenge and Registry (VCR) was specifically endorsed as a national program element.[26]

VCR, like other voluntary programs, has no legislative basis; but it does have legal foundations. VCR is an incorporated, not-for-profit entity established to manage the program. In carrying out its operations, it is subject to legal corporate governance principles, and in its operations, including assessment and credit for voluntary programs submitted, it must treat its member participants in a procedurally fair manner (Lucas, 2000).

In addition, voluntary programs may be given legal force under regulatory legislation if they are incorporated as terms and conditions of regulatory approvals. If this is done, in principle, voluntary commitments become enforceable through administrative orders under statutory powers and even, subject to some difficult evidentiary issues, through prosecution for regulatory offences based on failure to comply with approvals.

An example is the Alberta Energy and Utilities Board's February 2002 approval of TransAlta Energy Corporation's 900 MW Keephills Coal Fired Power Plant Expansion.[27] In its application, TransAlta made a commitment to offset emissions from the plant based on the CO_2 emissions from a combine-cycle natural gas facility of the same capacity. The Board specifically directed TransAlta to "fulfill its commitment in that regard." It further directed that these offsets be adjusted so that they correspond to any reasonable future changes in emission standards for such plants, and recommended the use of a third-party audit process to verify the offsets.

Liability Issues

Administrative law liability based on procedurally unfair or substantively unreasonable decisions is referred to above in relation to voluntary program schemes. Potential non-statutory liability associated with particular GHG reduction measures also requires attention. One issue concerns potential liability to resource rights owners and other affected citizens as a result of injection of carbon dioxide (CO_2) into underground formations, a technique that is being piloted in western Canada.

First, because the law of underground storage rights is unclear, care will have to be taken in securing rights to inject CO_2 into depleted natural gas reservoirs. Though at common law the mineral rights owner also owned the space occupied, where title to particular substances is divided, it is necessary to attempt to determine the intention of the granting instruments (Bankes and Bennett Jones, 2002). In Alberta, statutory provisions authorize the Energy and Utilities Board to regulate the underground storage of natural gas (Alberta, 1972, ss. 38, 39). The definition of "gas" may be wide enough to include CO_2 as a constituent of raw gas or other hydrocarbons (Alberta, 1972, s. 1[i][k][y]).

A second issue is liability for damage to the environment and to human health should injected CO_2 migrate and escape. In principle, owners of the substance, along with those responsible for injection and monitoring, may be liable on the basis of strict liability tort theory for damage resulting from escape of a "dangerous substance."[28] However, the probability of foreseeing such harm would be a relevant factor. Negligence theory requires intention, a more difficult element to establish, but negligence may be a basis for liability.

Could GHG emitters be liable for damage caused by consequential climate change or severe weather events? Could governments or private parties be liable for damage attributable to adaptive measures such as switching to genetically modified crops? Both are theoretical possibilities, but storms and other extraordinary natural phenomena may provide an act of God defence if these events and their consequences could not be foreseen (Linden, 2001, p. 510). However, even given rapidly improving predictive capabilities, what can reasonably be considered foreseeable? Even more important, in both emission and adaptive action cases, how can cause be established and a particular damage proven to the balance of probabilities standard? Of course, such liability could be removed or limited by a statute of the relevant federal or provincial jurisdiction, as the federal government did in the case of nuclear liability (Canada, 1975). However, such legislation would raise obvious public concerns about fairness, equity, and public welfare.

Demonstrating proof of damage is the fundamental difficulty in establishing any liability related to climate change. For example, corporate participants in the now-defunct Global Climate Coalition lobby risked little other than public credibility since it is unlikely that anti-climate change positions endorsed by the group amounted to fraud or defamation, and what specific damage could be proven by opposing parties? There is a better possibility of corporate shareholders establishing damage to their interests in actions against corporations that commit corporate resources to voluntary GHG reduction initiatives.

Conclusion

The Canadian legal context for climate change includes international and national law elements. At the international level, the legal regime continues to evolve. Ratification by a number of countries sufficient to bring the Kyoto Protocol into force as an international agreement will be a critical step. In the meantime, conventions of the countries party to the Protocol continue to develop the operating rules and processes necessary to achieve the fundamental objectives of the Protocol.

At the level of Canadian national law, ratification binds Canada as a nation state; but binding Canadian corporations and individual citizens is a matter of national legislation. This national legislation must respect the constitutional division of legislative powers, and divided constitutional jurisdiction over the range of actions and instruments available for GHG reduction purposes creates a major potential impediment to timely and effective implementation. Federal-provincial agreement in some form appears to be essential.

Otherwise, framework law and enabling legislation exist at both federal and provincial levels that can be used to establish enforceable standards, subsidy and incentive programs, and emission trading schemes. Though some new legislation and amendment of existing legislation may be required, the capacity for legal adaption is high provided that the problem of constitutional jurisdiction can be managed cooperatively.

Notes

1 When it comes into force 90 days after it is ratified by 55 countries representing in total 55% of 1990 GHG emissions (Art. 25.1).
2 At the COP 7 in Marrakech, November 2001, based on Decision 5/CP. 6, "Implementation of the Buenos Aires Plan of Action."
3 As required by the Vienna Convention on the Law of Treaties, open for signature 23 May 1969, 8 ILM 679.
4 4 December 1992.
5 The process or "Berlin Mandate" was established by Decision 1/CP. 1.
6 Through the Ad Hoc Group on the Berlin Mandate, established at COP 1.

7 Fifty-five countries party to the UNFCCC accounting in total for at least 55% of 1990 emissions of Annex 1 parties (Art. 25).
8 COP 1 to COP 7, the latter held in Marrakech, November 2001.
9 Alberta's *Climate Change and Emission Management Act*, passed in late 2003, creates the enabling authority for these matters to be addressed by regulations, but as of March 2004 was not proclaimed in force.
10 *A-G Canada v. A-G Ontario (Labour Conventions Case)*, [1937] A.C. 326 (P.C.). The Canadian National Implementation Strategy on Climate Change, October 2000, the "recognition that jurisdictions have authority to develop specific programs and flexibility to reflect their unique circumstances" (p. 12).
11 *R. v. Crown Zellerbach Ltd.*, [1988] 1 S.C.R. 401.
12 *Pronto Uranium Mines v. Ontario Labour Relations Board*, (1956) O.R. 862 (Ont. H.C.).
13 *R. v. Hydro Quebec*, [1997] 3 S.C.R. 213.
14 Particularly the federal taxation power discussed below.
15 Although a reference to a provincial Court of Appeal for an advisory opinion can be made based on hypothetical facts, as in *Re Exported Natural Gas Tax (1982)*, [1982] S.C.R. 1004, the range of potential federal legislative instruments and approaches is so broad that design of a relevant hypothetical is difficult.
16 A factor weighing in favour of federal peace, order, and good government jurisdiction: *R. v. Crown Zellerbach Ltd.*, [1988] 1 S.C.R. 401.
17 *R. v. Crown Zellerbach Ltd., supra.*
18 *114957 Canada Ltée (Spraytech, Société d'arrosage) v. Hudson* [*Town*] [*Spraytech case*], [2001] 2 S.C.R. 24.
19 Substances can be regulated only after both being assessed and found toxic or capable of becoming toxic, and then being added by Cabinet order to the list of toxic substances in Schedule 1.
20 In s. 64 of Part 5, Controlling Toxic Substances, "Toxic" is defined as follows: "a substance is toxic if it is entering or may enter the environment in a quantity or concentration or under conditions that
(a) have or may have an immediate or long-term harmful effect on the environment or its biological diversity;
(b) constitute or may constitute a danger to the environment on which life depends; or
(c) constitute or may constitute a danger in Canada to human life or health.
21 Regulations can be made by Cabinet only after toxicity assessment, ministerial recommendation, and listing of a substance in Schedule 1, and an objection to any of these decisions will trigger a Board of Review proceeding (s. 333).
22 E.g., Alberta Energy and Utilities Board (EUB) Decision 2002-014, TransAlta Corporation, 900-MW Keephills Power Plant Expansion, Application No. 2001200, February 2002 (69, 72); EUB Decision 99-2. Shell Canada Limited. Muskeg River Mine Project, Application No. 970588, February 1999 (30).
23 Since sections 2 (purposes) and 3 (public interest factors) of the Energy Resources Conservation Act, RSA 2000, c. E-10, concern development and conservation of Alberta energy resources and the environmental effects in Alberta, global warming actions may, without specific statutory authority, be considered legally irrelevant to the Board's exercise of its discretionary powers.

24 *Re Rizzo and Rizzo Shoes Ltd.*, [1998] 1 S.C.R. 27.
25 *Spraytech* case.
26 VCR Inc. <http://www.vcr-mvr.ca> was prominent in the 1998 joint ministerial statement on the Kyoto Protocol: Joint Meeting of Energy and Environment Ministers, Record of Decision, Summary of Decisions, No. 8, The Voluntary Challenge and Registry Inc., 20 October 1998.
27 Alberta Energy and Utilities Board, Decision, *supra* at 69: TransAlta Energy Corporation. 900-MW Keephills Power Plant Expansion, February 2002.
28 *Rylands v. Fletcher* (1868), L.R. 3 H.L. 330; *Cambridge Water Company v. Eastern Counties Leather*, [1994] 1 All E.R. 53 (U.K. H.L.).

References

Adam, M. (2002). Mayors support Ottawa on Kyoto. *Calgary Herald*, 18 February 2002, A-1.

Alberta. (1957). Oil and Gas Conservation Act, R.S.A. 2000, c. O-6.

———. (1988). Small Power Research and Development Act, R.S.A. 2000, c. S-9. Small Power Research and Development Regulation, Alta. Reg. 336/88.

———. (1992). Environmental Protection and Enhancement Act (EPEA), R.S.A. 2000, c. E-12.

Bachelder, A. (2000). Using Credit Trading to Reduce Greenhouse Gas Emissions. *Journal of Environmental Law and Practice* 9: 281–98.

Bankes, N. (1991). Shaping the Future or Meeting the Challenge? The Federal Constitutional Proposals and Global Warming. *Resources* 36: 1.

Bankes, N., and Bennett Jones, eds. (2002). *Canadian Oil and Gas*. Vol. 1, Toronto: Butterworths.

Bueckert, D. (2002). Kyoto Plan Stays, Ottawa Insists. *Calgary Herald*, 19 February 2002, A-4.

Canada. (1969). Nuclear Liability Act, R.S.C. 2000, c. N-28.

———. (1981). Motor Vehicle Fuel Consumption Standards Act, S.C. 1981–82, c. 113.

———. (1992). Energy Efficiency Act, S.C. 1992, c. 36.

———. (1999). Canadian Environmental Protection Act, (CEPA), 1999, S.C. 1999, c. 33.

Canada Gazette Part II. (1995). Energy Efficiency Regulations, SOR/94-651, 97-529; Ozone Depleting Substances Regulations, SOR/95-576.

Castrilli, J. (1999). *Legal Authority for Emissions Trading in Canada: Legislative Authority to Implement a Domestic Emissions Trading System*. Ottawa: National Round Table on the Environment and the Economy.

Constitution Act. (1867). (U.K.), 30 & 31 Vic., c. 3 [formerly British North America Act, 1867].

Dunham, L. (2002). Proposed Canadian Domestic Greenhouse Gas Emission Trading Programs: Are They Compliant with International Trade Law? Faculty of Law, University of Calgary.

Environment Canada. (1995). *National Action Program on Climate Change, Strategic Framework of the Action Program*. Ottawa: Environment Canada.

Ferguson, E. (2002). Klein Cools Off on Kyoto Challenge. *Calgary Herald*, 28 February, 2002, A-1.

Grimeuld, D. (2001). An Overview of the Policy and Legal Aspects of the International Climate Change Regime. Pts 1 and 2. *Environmental Liability* 2: 39–52; 3: 95–126.

Hogg, P. (1997). *Constitutional Law of Canada*, 4th ed. Toronto: Carswell.

Jaimet, K. (2002). Chretien Rejects Kyoto Warnings. *Calgary Herald.* 28 February 2002, A-1.

Kyoto Protocol to the United Nations Convention on Climate Change. (1997). UN Doc. FCCC/CP/1997/L.7/Add. 1.

Linden, A. (2001). *Canadian Tort Law.* 7th ed. Toronto: Butterworths.

Lucas, A. (2000). Voluntary Initiatives for Greenhouse Gas Reduction: The Legal Implications. *Journal of Environmental Law and Practice* 10: 89–104.

Mimms, J. (2001). The Marrakech Accords on Climate Change: Recent Progress on the Kyoto Protocol. *Environmental Liability* 6: 275–85.

Montreal Protocol. (1987). *On Substances that Deplete the Ozone Layer, 1987.* 1 January 1989, amended 1990, 27 I.L.M. 868.

North American Free Trade Agreement (NAFTA) (1992). 7 October 1992, 32 I.L.M. 289, 605.

Petsonk, A. (1999). The Kyoto Protocol and the WTO: Integrating Greenhouse Gas Emissions Allowance Trading into the Global Marketplace. *Duke Environmental Law and Policy Forum* 10: 185–220.

Rolfe, C. (1998). *Turning Down the Heat: Emissions Trading and Canadian Implementation of the Kyoto Protocol.* Vancouver: West Coast Environmental Law Research Foundation.

Stewart, R., et al., (1996). Legal Issues Presented by an International Greenhouse Gas Trading System. UN Conference on Trade and Development Pub. No. UNCTAD/GDS/GFSB, Misc. 1. Geneva: UNCTAD.

Transport Canada. (1996). *Transportation in Canada.* Ottawa: Transport Canada.

Transportation Issue Table. (1998*). National Climate Change Process; Foundation Paper on Climate Change; Transportation Sector.*

Tellus Institute. (2002). *The Bottom Line on Kyoto.* David Suzuki Foundation and World Wildlife Fund. <www.davidsuzuki.org/files/kyotoreport.pdf>.

United Nations Framework Convention on Climate Change (UNFCCC). (1994). 9 May 1992. <http://www.unfccc.int/resource/conv/index.html>.

Vienna Convention on the Law of Treaties. (1969). In force 27 January 1980. 1155 U.N.T.S. 331.

World Trade Organization Agreement (WTO). (1994). 33 I.L.M. 1125. <http://www.wto.org/english/docse/legal/finale.htm>.

PART III

Hard Choices

A Canadian Policy Chronicle

James P. Bruce and Doug Russell

A Little History from a Canadian Perspective

CLIMATE CHANGE IS OFTEN CALLED a "science-driven" issue. Indeed, it was the scientific community that ensured its consideration globally and nationally, but the issue has now moved well beyond climate science. As noted in chapter 1, scientists in the nineteenth century, such as Fourier in France and Arrhenius in Sweden, were well aware of the so-called "greenhouse effect." In 1898, well after the beginning of the industrial revolution, Arrhenius noted that the continued burning of fossil fuels would increase atmospheric carbon dioxide (CO_2) concentrations. He even made an estimate of the mean global temperature increase that might occur, which is reasonably close to current estimates.

However, aside from a few individual scientists, no concerted global scientific effort was made to pursue the subject until the First World Climate Conference in 1979 in Geneva. This Conference was sponsored by the World Meteorological Organization, and one of the key organizers was Kenneth Hare, then of McGill University. The Conference addressed many climate-related issues, such as the modes of natural variability and the economic uses of climatic data, but it also raised a warning flag about the potential impact of human activities on the world's climate. The Conference was the springboard to launch the World Climate Research Program.

About the same time, the US National Academy of Science released reports on CO_2 and climate change. In Canada, Assistant Deputy Minister Arthur Collin of the Atmospheric Environment Service formed the Canadian Climate Centre, bringing together the climate expertise of his organization. He also established the Canadian Climate Program Board, consisting of experts from a number of departments and agencies from both within the government and elsewhere, to coordinate Canadian activities on climate science and its policy implications. Kenneth Hare was appointed the first Chair of the Canadian Climate Program Board.

Hare and Collin were just the first of many Canadians to contribute substantially to international science and policy on climate and climate change. A major international conference that first began calls for action to reduce greenhouse gas emissions was held in Villach, Austria, in 1985, and was chaired by Canadian James P. Bruce. That conference had before it a report of a Panel of the International Council of Scientific Unions (ICSU), which made it clear that other greenhouse gases, such as methane and nitrous oxide, are also increasing rapidly in the atmosphere and are greatly augmenting the effects of CO_2. The Villach Conference sent a message to the world that much costly infrastructure and many human activities are planned on the assumption that past climate is a reliable guide to the future. The Conference concluded, "This is no longer a good assumption" (WMO, 1986).

To follow up the Villach 1985 results, the three Conference sponsors, the World Meteorological Organization (WMO), the United Nations Environment Program (UNEP), and the International Council of Scientific Unions (ICSU), formed an Advisory Group on Greenhouse Gases of seven eminent scientists, chaired by Kenneth Hare. The panel was to advise the three sponsoring organizations, and through them the world's governments, about the climatic implications of continued emissions of greenhouse gases. They issued several valuable reports.

In 1986, the Bruntland Commission (World Commission on Environment and Development) with Canadian James MacNeill as Secretary-General, held influential public hearings in Ottawa, where human impacts on climate were discussed. Then, in 1988, Canada hosted a landmark international conference on "The Changing Atmosphere—Implications for Global Security." This conference called on developed countries to commit themselves to CO_2 emission reductions of 20% from 1988 levels by 2005 (WMO, 1988).

By this time some governments, especially the US, had become uneasy about leaving international pronouncements on the science of climate change and its implications to the Advisory Group on Greenhouse Gases, the small panel of experts. These governments persuaded the WMO and later the UNEP that a much more inclusive and comprehensive science assessment process was needed, one in which governments could influence, but not dictate, the outcome. Thus, the Intergovernmental Panel on Climate Change (IPCC) was formed in November 1988 by the WMO and the UNEP (Bruce, 2001).

In Canada, the House of Commons Standing Committee on the Environment in 1990 and 1991 called on Canada to adopt the Toronto Conference target (20% reduction by 2005) as an interim objective. The then Minister of Environment, Lucien Bouchard, at the May 1990 UN Conference on Sustainable Development in Bergen committed Canada to stabilize CO_2 emissions at 1990 levels by 2000. Also in 1990, the federal government adopted a comprehensive "Green Plan for a Healthy Environment" that reinforced the commitment to stabilize emissions of CO_2 and other greenhouse gases. The Green Plan also provided funding for a number of related initiatives to climate change, including an increase in climate change research, programs to encourage efficient use of energy, and a framework for Canada's overall climate change response strategy.

IPCC's First Assessment Report was released in 1990 and was considered in November 1990 by the Second World Climate Conference in Geneva (also organized by a Canadian, Howard L. Ferguson) (Jager and Ferguson, [1991]). At this global event, political leaders from Margaret Thatcher to the President of the Maldives called for an international agreement to curb emissions of greenhouse gases. The UN set up an international negotiating process in December 1990, and the first round of negotiations was hosted in February 1991 by the United States.

In the negotiations that followed, Canada emphasized a net approach to all greenhouse gases, to include both sources and sinks. During the course of negotiations, realization of the economic importance of the issue led to changes in the composition of the Canadian delegation. It was expanded to include not only scientific leaders in the Atmospheric Environment Service, Environment Canada, and diplomats from the Department of Foreign Affairs, but also other departments and environmental and business organizations dealing with energy and the economy. Elizabeth Dowdeswell of Environment Canada was co-chair of the negotiations that established the legal framework for the reporting, review, and compliance provisions. These remain in force today. By 1992, an agreement had been drafted and accepted at the World Conference on Environment and Development (Rio de Janeiro) with another Canadian—Maurice Strong—at the helm. This agreement is the United Nations Framework Convention on Climate Change (UNFCCC), which was subsequently ratified by 186 countries,[1] including Canada and the US. The UNFCCC entered into force in March 1994; its ultimate objective to "achieve ... stabilization of greenhouse gas concentrations in the atmosphere at a level that would prevent dangerous anthropogenic interference with the climate system."[2] It called on countries to reduce greenhouse gas emissions and to increase and protect carbon removals by sinks (vegetation, soils). Industrialized countries were required to report regularly on efforts "aimed at" returning emissions to 1990 levels, by the "end of the decade," i.e., 2000. The UNFCCC also contained provisions for governments to cooperate in research, systematic observations of the climate system, and measures to adapt to the changing climate.

Post-Rio Negotiations

In 1995, at their first official meeting in Berlin, parties to the UNFCCC reviewed the adequacy of the commitments in the convention and decided that the commitments would not be sufficient to stabilize global concentrations of greenhouse gases. The parties adopted the Berlin Mandate, which set the stage for two and a half years of negotiations on firmer emissions reduction commitments and the eventual 1997 Kyoto Protocol. During the period of these negotiations, a Second Assessment Report was released by the IPCC (1996), covering the economic and social dimensions of climate change as well as the natural sciences. The report noted that human activities were indeed already affecting the global climate, and that many measures exist to reduce emissions and increase sinks. A number of these measures would have low or negative costs or net benefits to national economies. Driven in part by the consensus view influenced by IPCC, and mindful of increasing evidence of climate change, national representatives gathered in Kyoto, Japan, and in December 1997 adopted a Protocol to the Convention. The Kyoto Protocol contains firmer commitments by industrialized countries to reduce net emissions of greenhouse gases by the period 2008 to 2012. Canada's target is a 6% reduction below 1990 levels during that commitment period (see chap. 3).

Following adoption of the Kyoto Protocol in December 1997, international negotiations focused on fleshing out the details of the Protocol in a manner that would allow countries the ability to make their ratification decisions. Many of the concepts agreed to in Kyoto are groundbreaking for international environmental agreements, in particular, the emergence of a global market for greenhouse gas emissions (Bernstein and Gore, 2001). Numerous complex technical and political issues had been left unresolved at Kyoto and the following year, at the fourth Conference of the Parties (COP 4) in Argentina, negotiators hammered out the Buenos Aires Plan of Action. The plan laid out a two-year schedule to resolve outstanding issues surrounding the rules on how the Kyoto Protocol would be implemented. These included procedures and rules surrounding sinks (land use, land use change, and forestry), the Kyoto mechanisms (the Clean Development Mechanism, Joint Implementation and International Emissions Trading), reporting requirements, the Protocol's compliance regime, and other issues.

By November 2000, when Ministers met at COP 6 in The Hague, many thought the stage was set for final resolution. Canada entered COP 6 with a mandate to seek, among other things, credit under the Kyoto mechanisms for nuclear power projects, an increased allotment for managed forests, and no limits on the amount of emissions credits that could be purchased offshore for application towards Canada's Kyoto target. Following long hours and all-night negotiations, talks broke down in acrimonious and public fashion. In May of the following year, the US Bush administration withdrew support for Kyoto but indicated that it would continue to meet its commitments under the UNFCCC. This withdrawal opened the door for Canada to follow suit, but a political decision was made to not side with the US, and Canadian negotiators were instructed to resume talks in July 2001 in Bonn. With the US now officially out of the Protocol negotiations, and with many European and developing countries fearing that the Protocol hung in the balance, Canada and Russia were able to negotiate additional concessions (short of credit for nuclear power in the first commitment period), particularly related to sinks. Somewhat to the surprise of many observers, a deal was struck—one that paved the way in November of 2001 for the approval of the Marrakech Accords at COP 7. The Accords provide detailed legal guidance on the rules of how Kyoto will work and have cleared the way for many countries, including the European Union members and Japan, to ratify the Kyoto Protocol.[3]

Canada has now also ratified the Kyoto Protocol, despite competitiveness concerns arising from the US decision to withdraw.[4] US non-participation is a serious problem since that country is responsible for about one quarter of global greenhouse gas emissions. The key additional country that needs to ratify the protocol is Russia, which has delayed a decision pending the assurance of a developed-country market to which emission credits can be sold.[5] Should Russia ratify it, the Kyoto Protocol will come into force.

Canadian Emission Reduction Initiatives

In Canada, consensus on an emission reduction strategy has proved difficult even within the federal government, with seriously conflicting views between Natural Resources Canada, which is responsible for federal energy policy, and Environment Canada, whose mandate is environment policy. Canadian policy development has been influenced by perceptions, real and imagined, about Canada's "national circumstances," and complicated by the division of powers between the federal and provincial governments. Canada has about 0.5% of the world's people but emits 2% of the anthropogenic greenhouse gases. Among larger countries, Canada's greenhouse gas emissions per capita are second only to those of the US. Emissions vary greatly, ranging from 640 kt CO_2-equivalent per year[6] for the Yukon to 199,000 kt CO_2-equivalent for Alberta, for a national total of 682,000 kt CO_2-equivalent (1997 data) (Environment Canada, 1999).

The federal, provincial, and territorial energy and environmental ministers are jointly responsible for the development of Canada's response to climate change. They are also consulted on Canada's negotiating positions for international climate change talks (Joint Ministerial Meetings).

Prior to the Kyoto meeting in 1997 (COP 3) the G8 leaders met in Denver and agreed on "meaningful and realistic targets to be achieved by 2010." The Energy and Environment Ministers (JMM) Council had debated Canada's position and would only go as far as endorsing a goal of returning emissions to 1990 levels by 2010. However, the federal government had announced just before the December 1997 Kyoto session that it would accept a −3% target by 2010 and a further −5% by 2015 (Canada, 1997). In the end, in Kyoto, Canadian negotiators responded to a surprise Clinton administration move to commit to a 7% reduction, and went further than agreements reached by the Council in accepting a 6% reduction target below 1990 levels by the 2008–2012 commitment period.

Most provincial and territorial governments and many in the business community objected strongly to the target, resulting in ongoing acrimonious debate in post-Kyoto discussions about the methods and costs of achieving the target and whether Canada should ratify the Protocol. It should be noted that between 1990 and 1997 emissions in Canada had risen by 14%, and little action other than the Voluntary Challenge had been taken to address provisions of the Framework Convention of 1992. In fact, these emissions have continued to rise since 1997 and "business as usual" projections to the 2008–2012 period suggests emissions of 26% or more above the 1990 level, turning the 6% reduction target into a need for a 32% reduction in projected net emissions (National Climate Change Process, June 2002). Each year that goes by without substantive action makes the target more difficult to achieve (Russell and Toner, 1999; Bernstein and Gore, 2001).

Following 1997 and Canada's signing of the Protocol (29 April 1998), the Energy and Environment Ministers Council (JMM) agreed to establish a series of sectoral "Tables" to analyze options to help meet Canada's Kyoto emission target. The 1998 federal budget provided $150 million over three years for a Climate Change Action Fund. This budget made provisions to fund the Issues Tables as well as to support science, studies of impacts and adaptation, technology development, and public outreach. The 16 Tables included agriculture, transportation, electricity, municipalities, and a number of others, including some cross-sectoral Tables on carbon sinks, economic analysis, market mechanisms, credit for early action, and technology. These Tables were, in effect, advisory panels, with a broad representation from industry, universities, government, and non-governmental agencies, some with environmental mandates. The federal and provincial governments established a National Climate Change Secretariat to coordinate the analytical work.[7] A parallel federal Secretariat was also established, reporting to the ministers of environment and natural resources.

The Tables were structured to work on a consensus basis with strong industrial and environmental non-government organization (ENGO) interests at each Table. It was perhaps inevitable that what could be achieved by consensus fell short of providing the means of closing Canada's Kyoto gap. While numerous specific measures were analyzed, no coherent, integrated plan emerged from compiling results from the Tables; nor have they provided guidance to ensure that no one region of the country bears a disproportionate share of the economic burden of reaching the Kyoto target, as the prime minister has promised. Meeting this objective will require tough political decisions, since it is clear that some provinces and industries would gain economically from an emissions reduction program based on reduced use of fossil fuels, and others would lose.

To date, federal and provincial ministers have not agreed on a course of action. Several provincial premiers spoke out against ratification of the Kyoto Protocol, as have several business organizations. The Canadian Council of CEOs released in June 2002 an alternative plan based on technology development and rate-based targets.[8] Provincial objection is related to the potential of lost revenue from fossil-fuel reduction and concerns that US non-participation would create short-term competitive advantages for US firms. However, numerous economic studies have shown[9] that the lack of demand for international emissions credits (arising from the US being out of the market) would reduce the costs of purchasing credits, and therefore the costs of compliance for other countries from 2 to 10 times (see note 9). Furthermore, the US is not standing still. With a number of multipollutant bills being drafted or considered in state legislatures, there is pressure on US industry (including vehicle manufacturers), to address greenhouse gas emissions. However, many Canadian industries remain concerned that the opposition to Kyoto in the White House creates a real problem for Canadian industry.

Governments in Canada, in order to stimulate economic growth, have provided financial support such as fuel-tax credits and direct subsidies to many industries. For fossil fuels in Canada, estimates have run as high as $5.9 billion per year (Myers and Kent, 2001). Other industrialized countries such as the US and Australia have granted similar subsidies. These subsidies, while perhaps beneficial in the short term, have arguably inhibited the development of alternative energy sources, which are vital to put Canada and other countries on a path to meet international climate change commitments. Another factor inhibiting actions in North America is that gasoline and fossil-fuel prices paid by consumers are among the lowest in the world, and there is strong objection to changing. This makes adjustments to consumer behaviour difficult. However, Canadian drivers polled by the Canadian Automobile Association see a mandatory or negotiated tightening of fuel efficiency standards as the best way of dealing with vehicle emissions.

To 2002, Canada's official responses concerning emission limitations and increasing sinks, under both the UNFCCC and the Kyoto Protocol, have been long on consultation and rhetoric, and short on significant action. The Voluntary Challenge has been shown by independent analysis to have achieved few emission reductions (Pembina Institute, 1997). Nevertheless, there are some encouraging signs. A number of Canadian cities and towns—where 80% of Canadians live—are members of a 20% reduction club and have taken substantial action. For example, the city government of Toronto has cut its own greenhouse gas emissions by some 64% over the past decade through improvements in energy efficiency in buildings and vehicles, and by capturing and using methane from landfill sites for energy, all at little or no net cost to the city administration. The Federation of Canadian Municipalities, in marked contrast to the provincial premiers, called in March 2002 on the federal government to ratify the Kyoto Protocol. In addition, some Canadian corporations and multinational companies operating in Canada have voluntarily committed themselves to emission limitations. For example, Ontario Power Generation has committed to a stabilization of its emissions at 1990 levels for 2000 and beyond; Dupont has a target to reduce GHG emissions by 65% by 2010 from 1990 levels; TransAlta has pledged to achieve zero net GHG emissions from the company's Canadian operations by 2024; Shell (worldwide) has achieved its target of a 10% reduction of GHG emissions from 1990 levels by 2002; and Husky Oil has introduced major energy efficiency measures (Russell and Margolick, 2001).

In May 2002, the federal government released details and associated analysis for four options to achieve the Kyoto Protocol target. A series of meetings has been conducted across the country, seeking views on the potential permutations and combinations of the options. The most promising scenario involves setting mandatory targets for key industries and establishing an emissions trading system to reduce costs. It also includes federal responsibility to fund energy conservation and alternative energy measures, purchase of credits from abroad, and receipt of credit from the international community for export of cleaner energy (natural gas and hydro) to the US. The idea of credit for clean energy exports has been soundly criticized by Europe and others as an unworkable proposal.

Alberta, the major oil, gas and coal-producing province, advanced an alternative "made in Canada" option that stops well short of achieving the Kyoto targets. The Alberta plan parallels President Bush's Clean Skies Initiative and that of the Canadian Council of CEOs, emphasizing technological solutions, rate-based targets, and voluntary measures (see chap. 10).

These pre-ratification consultations took place in the midst of a very vigorous public relations war. Some members of the business community and Alberta claimed that meeting Kyoto targets would cripple the economy and cost 450,000 jobs (on the assumption that many factories would move to the US). Estimates of the economic impact, from various economic models with widely varying assumptions, range from a loss of $23 billion to a gain of $5 billion nationally per year. Environmental groups provided analyses to suggest that Canadian competitiveness would be enhanced through energy efficiency and renewable energy measures (Boustie et al., 2002). Much depends on the measures selected and how they are implemented. A resounding message that was heard across the country during the course of the consultations—from both the ENGO and business community—was that a decision was needed soon. Since ratification, the public relations war has subsided, although concrete steps towards emissions reduction have yet to be taken.

Canada's Response to Requirements for International Cooperation on Science and Adaptation

As noted, the UN Framework Convention on Climate Change, reinforced by the Kyoto Protocol, called upon governments to take a number of actions beyond reducing greenhouse gas emissions and enhancing sinks. Among these actions are regular reporting, including estimates of greenhouse gases emitted, and cooperation in science, in systematic observations, and in the development of adaptation strategies to cope with the impacts of changing climate and related sea-level rise. Much of the global scientific research has been coordinated through the World Climate Research Program (WCRP) led by WMO, the Intergovernmental Oceanographic Commission of UNESCO, and ICSU (now the International Council of Science). This program was designed by a Joint Scientific Committee drawn from the international organizations and for a considerable time chaired by another Canadian, Gordon A. McBean. Programs to complement the physical-science-based WCRP were also developed by the international scientific community, notably the International Geosphere Biosphere Program (IGBP) and the International Human Dimensions of Global Change Programs (IHDP). Their combined outputs, including research and data collection from Canada, have continued to provide an important part of the natural and social science results assessed by IPCC. At the same time, Canadians such as Ian Burton and Barry Smit took a lead in international studies of adaptation options and in techniques for assessing such options.

Canada has been active in all of these national and international cooperation activities, although its relatively low level of commitment to science and systematic observations has come under some criticism. Major scientific efforts have included BOREAS, a study of the role of the Boreal forest in the climate system; and MAGS, the Mackenzie Basin GEWEX Study. (GEWEX is the Global Energy and Water Cycle Experiment). Many adaptation studies have been conducted but only a few measures actually undertaken to date in Canada. One notable action was to allow for future sea-level rise in the design of the fixed link bridge over Northumberland Strait between New Brunswick and Prince Edward Island.

A special $100 million Canadian Climate Change Development Fund (CCCDF), administered by the Canadian International Development Agency (CIDA), is supporting projects in many countries to adapt to changing climate and sea-level rise, and to develop potential CDM projects. Programs are underway in the Caribbean, Nigeria, Brazil, Argentina, Colombia, India, China, and the Southwest Pacific, among other countries and regions. The federal government has also funded increases in domestic climate research, particularly through the new Canadian Fund for Climate and Atmospheric Sciences and through Canadian Climate Action Fund support of science and adaptation studies. A Foundation for Sustainable Technology Development has also been established and is now providing funding assistance to develop and demonstrate new technologies that promote sustainable development, with initial emphasis on technologies that mitigate, substitute, or sequester greenhouse gas emissions.

Conclusions

As the emission reduction policy debates continue, significant climate change impacts are appearing in Canada, consistent with climate models' projections of changes due to increasing levels of greenhouse gases in the atmosphere. Some of these observed and projected changes, and their impact in Canada, most severely in the Arctic and Prairies, are documented in chapter 4. It is recognized that Canada's emission reductions alone will have only a very minor impact on rates of climate change. However, it is also increasingly accepted that working together with other countries under international agreements, the Framework Convention and Kyoto Protocol, can begin to make a difference.

Kyoto is only a first small step towards stabilizing both greenhouse gas concentrations in the atmosphere and global climate. However, successful environmental agreements, on ozone layer depletion, acid rain, pollution of the Great Lakes, among other issues, have all started with a small step and expanded as needed. The first step is often the most difficult, and this is proving especially true for the issue of anthropogenic climate change since actions affect the energy base of the world's economy.

Notes

1 As of September, 2000.
2 Article 3, UN Framework Convention on Climate Change [1992].
3 For the Kyoto Protocol to become internationally binding law (i.e., enter into force), two conditions must be met: at least 55 countries need to ratify the Protocol (done) and the countries that have ratified must represent at least 55% of the developed countries' emissions. This second condition has not yet been met.
4 Canada ratified the Kyoto Protocol on 17 December 2002, after a Parliamentary vote.
5 Russia, along with other countries whose economies are in transition to a market-based economy, suffered a decline in economic production in the 1990s and, by virtue of the agreements related to Kyoto will be a net seller of carbon allowances for the first commitment period (2008–2012).
6 kt CO_2 equiv. is 1000 tonnes of CO_2 or other greenhouse gases with global warming potentials (expressed as CO_2 equivalents).
7 The National Secretariat until May 2002 was co-chaired by the province of Alberta and the federal government. Following the May 2002 meeting of min-

isters, which rejected Alberta's proposal to have its alternative plan included formally as part of the pre-ratification discussions, Alberta resigned its position as co-chair of the National Secretariat.

8 Targets that reduce the amount of GHG emitted per unit of production. These rate-based targets can contribute to reducing emissions intensity but may not reduce total emissions if production continues to rise.

9 Six economic models, including Norwegian (CICERO), Australian (ABARE), and US (MIT and Yale) models, show that with the US out of the Protocol the projected price of CO_2 emissions per tonne in 2010 would fall from a range of $15-44 US to as low as <$1-$23.

References

Bernstein, S., and C. Gore. (2001). Policy Implications of the Kyoto Protocol for Canada. *isuma* 2(4): 26-36.

Boustie, Sylvie, Matthew Bramely, and Marlo Raynolds. (2002). *How Ratifying the Kyoto Protocol will Benefit Canada's Competitiveness.* Ottawa: Pembina Institute.

Bruce, J.P. (2001). Intergovernmental Panel on Climate Change and the Role of Science in Policy. *isuma* 2(4): 11-16.

Canada. (1997). *The Canadian Position on Global Climate Change: The Canadian Position for Kyoto: Background.* Ottawa: Environment Canada.

Commissioner of the Environment and Sustainable Development. (2001). *Report to House of Commons.* Ottawa: Auditor General of Canada.

Environment Canada. (1999). *Canada's Greenhouse Gas Inventory: 1997 Emissions and Removals with Trends.* Ottawa: Environment Canada.

House of Commons, Standing Committee on Environment and Sustainable Development. (1990-1991). *Our Changing Atmosphere.* 3 parts. Ottawa: Standing Committee on Environment and Sustainable Development.

———. (1997). *Kyoto and Beyond: Meeting the Climate Change Challenge.* Ottawa: Standing Committee on Environment and Sustainable Development.

Intergovernmental Panel on Climate Change (IPCC). (1996). *Climate Change 1995.* 3 vols. Cambridge, UK: Cambridge University Press.

———. (2001). *Climate Change 2001.* 3 vols. Cambridge, UK: Cambridge University Press.

Jager, J. and H.L. Ferguson, eds. (1991). *Climate Change Science, Impacts and Policy.* Proceedings of the Second World Climate Conference, Geneva 1990. Cambridge, UK: Cambridge University Press.

Myers, N., and J. Kent. (2001). *Perverse Subsidies: How Tax Dollars Can Undercut the Environment and the Economy.* Washington, DC: Island Press.

National Climate Change Process, Analysis and Modelling Group. (1999). *Canada's Emissions Outlook: An Update.* Ottawa: Natural Resources Canada.

Pembina Institute. *Corporate Action on Climate Change, 1997: An Independent Review.* Ottawa: Pembina Institue for Appropriate Development.

Russell, D., and M. Margolick. (2001). *Corporate Greenhouse Gas Reduction Targets.* Arlington, VA: Pew Center on Global Climate Change.

Russell, D.J., and G. Toner. (1999). *Science and Policy When the Heat Is Rising.* Ottawa: Global Change Strategies International.

Tackling Climate Change: Partnership Marketing Supplement. (2002). *Globe and Mail,* 16 March 2002.

WMO/UNEP/Government of Canada. (1988). *Conference Proceedings: The Changing Atmosphere.* Pub. 710. Geneva: WMO.

WMO/UNEP/ICSU. (1986). *Report of the International Conference on Assessment of the Role of Carbon Dioxide and other Greenhouse Gases in Climate Variations and Assorted Impacts.* Villach, Austria, 9–15 October 1985. Pub. 661. Geneva: WMO.

Beyond Kyoto

Gordon S. Smith and David G. Victor

Introduction

CLIMATE CHANGE IS ONE OF THE MOST IMPORTANT and vexing global challenges facing the world. Increasing numbers of people in Canada and around the world believe it is essential to find ways to control the buildup of greenhouse gases (GHGS) in the atmosphere. However, increasing economic competition means that few nations are willing to pass regulations to control emissions without a global framework to ensure that the costs are shared equitably. Moreover, effective international cooperation to address climate change requires that the measures taken actually reduce GHG levels and not just give the illusion of doing so.

We argue that the main impediment to action is the lack of a viable architecture for international cooperation—an architecture that both achieves results and assures that the playing field is not tilted against some nations. The previous chapter has described the process leading from the 1992 Earth Summit in Rio, to the United Nations Framework Convention on Climate Change (UNFCCC), to the general agreement reached amongst a number of countries in Kyoto five years later. The Kyoto agreement, which required significant further definition of the rules and, of course, ratification, has been the source of much debate during the ensuing years. At the seventh Conference of the Parties (COP 7) in Marrakech in the fall of 2001, governments agreed on nearly all the final details for the Kyoto Protocol; yet, at this writing, some states are facing the choice of whether and when to ratify. Difficult negotiations remain, as well as ongoing debates regarding the Kyoto agreement's long-term effectiveness in reducing GHGS, mechanisms to achieve commitments, equity issues, and economic costs.

The United States, which accounts for about one-quarter of the world's emissions, has indicated clearly that it rejects the Kyoto framework and will neither ratify the Kyoto commitments nor implement alternative commitments that would have a similar effect. Australia has also indicated that it will not ratify. Non-ratification by major industrialized countries presents a major problem to the success of the Protocol, which must be an *international* agreement to work effectively. Despite a considerable body of literature detailing the objectives and possible alternatives, there has been little clear thinking about its possible resolution.

After the Marrakech conference, Canada faced a set of unattractive policy choices within the Kyoto framework. Down one path, Canada could have simply followed the United States and withdrawn, claiming that it could not implement commitments that its largest trading partner had rejected. Such a path would have been politically costly, both nationally and internationally, possibly harming Canada's relations with the European Union (EU) and Japan. Most Canadians are concerned about climate change and wanted to see some effective action; in Canadian political debate, such action became equated with ratification of Kyoto.

Down the alternative path, Canada ratified the treaty and followed the European Union and Japan in an effort to bring the treaty into force. That approach was both politically and economically attractive because the US will not be competing for emissions credits; these credits constitute a major component of the Protocol, allowing nations time to implement stricter regulations to control national emissions while fulfilling Kyoto objectives to decrease global emissions (see chap. 10 for more information on emissions trading). Although the government of Alberta, Canada's largest petroleum extraction and refining province, and some in the business community—particularly in the energy-producing sector—have asserted that the economic costs of implementing Kyoto commitments would be high for Canada, these claims have not been proven. It seems that dissenters were not taking into account the US exit, which has eliminated the main source of demand for Kyoto emission credits. Moreover, the EU is implementing its Kyoto obligations through an internal EU emission trading system that will probably limit or prevent the import of surplus credits. With the US out, low or zero demand for surplus permits in the EU, and Russia likely to ratify the treaty, prices for emission credits will be very low—the market will be flooded with surplus permits.[1] As well, accounting changes adopted in 2001 will make it easier for countries with substantial forests, such as Canada, to meet the Kyoto targets at low cost as credit is given for additional carbon sequestration through "sinks" (see chap. 5).

Nevertheless, ratification is only a partial solution because the Kyoto Protocol, as we explain here, is unlikely to have much impact on global warming. Current estimates suggest there will be a reduction of perhaps 1% in global atmospheric concentration of GHGs from currently projected levels in the event Kyoto comes into force and nations meet their target reductions. The US exit has exposed other severe weaknesses in the Kyoto framework. The Kyoto approach gave inadequate attention to the cost of implementation, which means that countries (such as the US) that find compliance costs to be unexpectedly high have little choice but to exit. In addition, the exit of key parties can have severe effects on the costs incurred by all other parties—another crippling vulnerability of the Kyoto system.

In this chapter, we explore how Canada might play a role that appreciates the political reality that it is unlikely that the Protocol will be scrapped given the huge diplomatic and political investment already made in the Kyoto machinery. At the same time, Canada should find an approach that allows it to exert maximum leverage in creating an amended framework that will continue where Kyoto itself leaves off. Nevertheless, the agreement will have little effect on emissions through the year 2012. Canada's strategy concerning Kyoto, instead, should focus on building a better architecture for the period after 2012. No other country has presented a viable vision beyond Kyoto, and Canada may have an important interest and unique role in doing that. Canada's special role is based, in part, on its ability to advance novel principles that move beyond the US rejections of Kyoto and acceptance of it in its current form by the EU, which has led to squabbles between the EU and the US.

Position of the United States

The United States is, by far, the largest emitter of greenhouse gases. It is therefore vital to understand the position of the US government because of its pivotal role in any international framework for controlling emissions and because of the special economic relationship between the US and Canada, which will strongly affect Canada's policy choices.

Initially, the Bush administration emphasized uncertainties about the science of global warming, but it has largely abandoned that line of critique. The administration now concedes that climate change is occurring and that policy responses are needed to reduce levels of GHGs. However, US critique of the Kyoto framework as the means to reduce GHGs has been ongoing and can be traced back to earlier administrations and the first draft of the Kyoto Protocol. Hence, the US position cannot be said to be a partisan one, for it has been maintained in both Democratic and Republican administrations. At no time since 1997 has it seemed that even a majority of US Senators—let alone the required two-thirds majority—would have voted to ratify the Protocol.

The US has several objections to the Kyoto framework. The current administration and the majority in Congress believe that the economic costs are not domestically acceptable and that implementation would have major political repercussions. Another objection is the exclusion of developing countries (see chap. 3 for a discussion on Annex I and Annex II parties and developing country exclusion). President Bush pointed to both of these objections in a letter to senators Hagel, Helms, Craig, and Roberts, stating, "I oppose the Kyoto Protocol because it exempts 80 percent of the world, including major population centers such as China and India, from compliance, and would cause serious harm to the US economy" (Office of the Press Secretary, 2001).

Although the exclusion of developing countries from the Protocol stems from the argument that it was the industrialized countries that caused the problem and therefore should act first, the objection that developing countries do not have to make commitments is important, both politically and practically. Annual emissions of carbon dioxide from developing countries are rising rapidly, are on track to reach twice the 1990 level by 2010, and will likely soar even higher beyond 2010 (Energy Information Administration, 2002). Eventually these emissions must be capped, but in the short term, the developing countries are wary when the industrialized world has not yet implemented meaningful caps on its own emissions. One way forward is already found inside the Kyoto Protocol and constitutes one of the key elements of the Protocol that must be preserved: the Clean Development Mechanism (CDM). Serious efforts under the CDM must be coupled with a political strategy in the US to ensure that this form of engagement is accepted as proper participation, for now, by developing countries.

The foundation for acceptance of this approach of building on the CDM in the US is already in place, as a growing number of leaders in Congress have publicly called for action to control emissions in the US even if developing countries do not formally embrace binding caps. This issue has been portrayed in the media as impassable and fundamental, but on close inspection, the US may not be nearly so insistent upon imposing binding short-term caps on developing countries as has been portrayed. Many observers cite the 95–0 vote on the 1997 Byrd-Hagel resolution, which contains brief but strong wording on the need for developing country emission caps; that resolution, however, was adopted prior to the Kyoto Protocol, in part as a signal to the rest of the world, in order to strengthen the US negotiating position. Many senators that signed it have since announced their support for US emission controls even without binding caps on developing countries.

Other US objections demand more serious diplomatic attention and possible amendment of the Kyoto Protocol. One objection is that the Kyoto targets are too precipitous for the US, in spite of the fact the previous administration agreed to them. The actual targeted cut for the US, given the growth in emissions since 1990, would be 30% (McKibbin et al., 1999). There would have been only seven years from the time the Bush administration took power to achieve this target by the beginning of the 2008-2012 commitment period. This would have required the retiring of large (and in the US government's views excessive) amounts of capital stock with huge and unnecessary costs. The US also believes that the Protocol and subsequent negotiations excessively restrict the use of carbon sequestration, although clearly progress was made in that direction at COP 6 in Bonn in July 2001, as will be discussed below.

Finally, the US is concerned that the Kyoto Protocol and its timetable for implementation would have left the country dangerously dependent on other nations to meet US emission targets. While emissions trading could have reduced the costs to the US by one-half, few countries other than Russia and other Eastern European nations have the excess capacity to sell parts of their allocations. There is even then no guarantee that there would be emissions to sell, as countries wishing to do so must meet measuring and monitoring requirements. The availability of emissions credits may not be known until 2007, creating too great a risk about the costs of meeting US emissions targets in the view of many American critics of Kyoto. In addition, even if Eastern European countries were in a position to sell part of their allocations, American and other purchasers would simply transfer many billions of dollars without achieving any meaningful GHG reductions. This was and is deemed unacceptable, a point to which we return later in this chapter.

Canada's Position

At Kyoto, Canada committed itself to reduce greenhouse gas emissions 6% below 1990 levels during the period between 2008 and 2012. Prior to actual ratification, the position of the Canadian government was that Kyoto should be ratified, but negotiations were required so that Canada would be allowed to implement the agreed emissions limits in an acceptable way. Above all, Canada argued that the Kyoto accounting rules must be changed to allow credits for the export of "clean energy," such as exports of gas and hydroelectricity.[2] This position attracted criticism on two fronts. First, the idea of clean energy exports has been on the Kyoto agenda before—and was soundly rejected. Most other countries—notably those in the EU—think that the time has already passed for clever adjustments to accounting rules to make compliance easier. Without these other nations assent, Canada will be unable to get its concept of credit for clean energy exports adopted.[3] Second, criticism of Kyoto appeared in the rising tide of objections in Canada from the private sector, especially energy firms, journalists, and Alberta government officials. They generally, although not universally, expressed concerns about the costs to Canada in terms of lost opportunities for growth and a disadvantaged competitive position with the US as the two economies become more and more integrated.

In the spring of 2002, the Canadian government produced "A Discussion Paper on Canada's Contribution to Addressing Climate Change." The paper noted that the US decision not to ratify "creates a unique situation for Canada" (Canada, 2002, p. 5). The Protocol requires roughly comparable levels of effort from Canada and the US, but the US withdrawal creates an imbalance that "raises competitiveness issues for Canada and Canadian industry that need to be addressed" (Canada, 2002, p. 5).

Outlining a general plan for meeting Kyoto commitments, the paper acknowledged that overall, "business-as-usual projections would see Canada's GHG emissions rise to approximately 809 Mt [megatonnes] by 2010. Canada's Kyoto target is 571 Mt by 2010, creating a 'gap' of about 240 Mt that must be addressed" (Government of Canada, 2002, p. 15). Credit for exports would reduce the total required to 170 Mt. The discussion paper envisions that "some 74 Mt of the [remaining] gap could be closed as a result of actions and credits from current policies and programs" (Canada, 2002, p. 15). Sinks would provide 24 Mt of the 74 Mt. It was calculated that a gap of 96 Mt would still remain to be filled.

The paper then outlined four options for closing the gap: (1) a broad domestic emissions trading system that would include essentially all greenhouse gas sources and sinks; (2) targeted measures; (3) a mixed approach, centred on a domestic emission trading system that included only large emitters of greenhouse gases, such as fossil-fuel-powered electric power plants; and (4) an adjusted mixed approach. Most observers quickly concluded that the number and mix of losers made the first three options not politically viable, and the government's preference for the fourth seemed quite clear. This would involve 55 Mt coming from domestic emissions trading (new reductions for 25 Mt, offsets for 20 Mt and private sector purchases of international permits for the remaining 10 Mt). Targeted measures in the areas of technology, strategic investments, and best practices would produce savings of an additional 25 Mt. Government purchases of international permits would close the remaining 16 Mt, although it was also suggested that the government could purchase additional international permits to reduce risk of up to 30 Mt. It was estimated that the reduction in GDP from this approach might be 1% over the next decade. The price of carbon will determine the attractiveness of the permit system. Moreover, targets will not be met if carbon is priced too high, making the system inequitable. Calculations in the Discussion Paper suggest that the price of carbon might be C$10 to C$50, with the former thought to be most likely, but the latter the result of a "worst case" scenario.

The government of Ralph Klein of Alberta remains adamantly opposed to the approach suggested by the federal government. Alberta produced its own approach, which would essentially involve a delay in meeting the targets. Consultations with other western premiers and the leaders of the territorial governments, led by Klein, were directed at having the Alberta plan also considered with the four federal options.

Although this political process is far from over, we believe that two conclusions can already be drawn, and these conclusions will frame Canada's possible responses to the Kyoto challenge. First, ratification of and compliance with Kyoto is not significant to Canada because it is possible to comply with Kyoto simply through accounting provisions, which would allow Canada to take credit for its growing forests and other land-use practices (sinks) and to offset any deficit—even an extremely large deficit—through acquisition of extremely inexpensive credits (emissions trading) from Russia and Ukraine. The transition to a market economy in parts of the former Soviet Union has inadvertently lowered emissions due to a decline in industrial production, leaving Russia and Ukraine in a position to sell credits. Of course, this approach will not reduce emissions, a point that has not escaped environmental groups in Canada. However, the availability of these excess Russian and Ukrainian credits—a by-product of the allocation of targets in Kyoto and the fact that the US has withdrawn and thus will not compete to purchase them—allows every nation to set its own price for joining. Countries that are willing to bear a large national burden (e.g., those of the EU) will restrict imports of these credits and allow the price inside their countries to rise. Countries that are less willing (e.g., Canada, if Alberta gets its way) can open the way for Russian sales of carbon permits and drive down the price of permits and the cost of compliance.

Second, the government of Canada had hinged its ratification of Kyoto on a diplomatic achievement—credit for clean energy exports. An insistence upon clean energy export credits may be *designed* to produce diplomatic deadlock and failure, which would offer the Canadian government some cover against the domestic political backlash that will follow if it fails to meet its targets. If Canada succeeds in gaining these credits—which is extremely unlikely—then it can meet the Kyoto mandate at extremely low (or nearly zero) cost. If it fails, then it can object to ratification on manufactured grounds for saving face and pinning the blame on others.

The real issue at stake for Canada does not lie in choosing among these hypothetical options. Rather, for Canada, the strategic question revolves around how the government leverages its domestic policies to affect not only the trajectory of emissions within Canada but also the international framework for controlling emissions of greenhouse gases. Canada should neither squander its leverage over international diplomacy with a failure-prone bid to obtain clean energy export credits nor should it be concerned about the first Kyoto target period (2008-2012); that time is nearly here and there is little that can be done to alter emissions trajectories. Instead, the motivating force in Canadian climate policy should be the period *after* Kyoto. Canadian domestic policy should be used to demonstrate a more viable international framework for action after 2012. To explain this, we first explore why the Kyoto approach has failed and is likely to fail in the future.

The Crux of Kyoto's Woes: Allocation

The particular situation of the US facing onerous targets with no politically realistic way to comply except through the purchase of "hot air" from Russia and Ukraine is not merely the result of a misguided ambition that led governments to agree to more than they could (politically) deliver. Rather, the situation reveals a fundamental flaw in the "architecture" of the Kyoto Protocol—what one might call a "textbook trading" system in which countries adopt absolute caps on emissions and then trade emission credits so that each can meet its cap and the collective effort is efficient. This approach has been used successfully *within* some countries, notably in the US system for trading sulphur dioxide emission credits. However, creating and allocating emission credits across countries is quite different, especially when the marginal cost of controlling emissions is highly uncertain.

Emissions vary with the level of economic output and technology, neither of which can be planned by governments according to exacting targets. Since governments cannot plan future emission levels, they cannot be certain they will comply with particular emission targets, such as those in the Kyoto Protocol. Because the cost of meeting a particular target varies across countries and over time in unpredictable ways, it would be inefficient to set fixed limits on the quantity of allowable emissions in advance and force each nation to meet its target entirely within its own borders: some countries and firms will find less costly ways to limit emissions than others. Emissions trading is one of the Kyoto Protocol's proposed solutions to these problems. A nation in surplus can sell to another in deficit, allowing both to comply while costing in aggregate less than if each were forced to implement emission controls entirely on its own. Since greenhouse gases mix worldwide and climate change is a global phenomenon, it does not matter geographically where (or even exactly when) emissions are reduced. Countries that find themselves in deficit with no way to meet their targets on their own can buy surplus from others. Emissions trading is the lynchpin of the Kyoto agreement.

The problem with trading is that emission permits must be allocated before trading can begin—the cards must be dealt before the game commences. Allocation is the source of endless mischief and difficulty, resulting in a regime that, in practice, is unstable and likely to be ineffective. One difficulty is caused by the fact that emission permits could be extremely valuable. For example, the emission targets adopted in Kyoto implied an allocation of emission permits worth over US$2 trillion, calculated on the basis that the US was inside the regime. That sum reflected only the 38 states that accepted limits on their emissions as part of the Kyoto accord. Expanding the system to include more nations—the major developing countries, in particular—would have increased the size of the pie that had to be allocated (Victor, 2001). The world has never attempted an allocation of this magnitude.

Allocation of the global atmosphere would occur under international law, and that task is particularly difficult for three closely related reasons. First, because of the pace of ratification, agreements on allocation must be struck many years before they actually take effect. The result is that the value of allocations is highly uncertain because it is impossible to predict, far in advance, the marginal cost of abatement and because the exact value depends on the behaviour of many other countries—in particular, the number of countries that decide to ratify. In this environment of value uncertainty, countries that are wary of over-promising will demand extra permits ("headroom") in case their emissions or costs are higher than expected. As the number of participating countries rises, ironically it becomes harder to reach a deal because each new country asserts an additional demand for headroom.

This problem is not yet fully evident because it will be most severe only as the system is expanded to include the countries that are most wary of incurring costly obligations—notably, the developing countries that have not yet been part of the negotiations to cap emissions. An inkling of this problem was evident, however, in the Kyoto allocations for Russia and Ukraine. On the one hand, it was important to involve these countries in the Kyoto deal because they offered low-cost opportunities for controlling emissions (as do the developing countries); but, on the other hand, it was hard to entice these nations in because they were unwilling to devote resources to controlling emissions. The deal struck in Kyoto gave these nations emission entitlements that they could never need in the 2008–2012 period. The collapse of their economies and the resulting closure of factories have resulted in current emissions that are well below the levels of 1990. They participate because selling their "hot air," as it has become known, lets them earn money—by some estimates US$80 billion for Russia and US$20 billion for Ukraine if all the other nations with Kyoto targets (especially the US) participate.[4] These emission credits do not represent actual reductions; they are a windfall that is the by-product of the political process of allocation. Russia and Ukraine could bank these credits for the future when emissions constraints are tighter, in which case the credits might actually represent a bona fide reduction in emissions. The Kyoto model, however, envisions setting emission allocations every five years, and thus banking would be politically unwise: large quantities of credits in the bank weaken future allocation negotiations because they indicate that past allocations have been too generous.

Second, allocation is difficult because uncertainty and trading during the allocation process make it likely that emission permits will not be allocated along lines of equal marginal cost—countries that have low abatement costs, such as Russia and most of the developing world, will probably get many more permits than they need because they are unwilling to devote large amounts of (or any) national resources to controlling emissions. They can credibly demand extra permits and threaten to exit the regime if they do not get what they want. Others that have high marginal costs, such as most of the advanced industrialized countries, nonetheless accept many fewer permits than they will need because they are under strong domestic pressure to do something about global warming. The latter are over a barrel and must agree if they want any action to reduce emissions—unless, as in the US today, they are pressured by interest groups that stress the importance of economic growth over the threat of global warming.

The result of these bargaining dynamics under uncertainty, however, is the big political liability from emissions trading: huge flows of money occur when the system is initiated before the market equilibrates (for example, the flows of tens of billions of dollars for Russia and Ukraine). Unless the allocation is nearly perfect, the flows will be large—much larger than current flows of development assistance. For comparison, the US$100 billion that might have gone to Russian and Ukraine under the Kyoto allocation is approximately twice the yearly flow of official development assistance to all developing countries combined. Some flow during equilibrium is tolerable as the necessary cost of getting other countries on board, but we have already seen from the experience with Russia and Ukraine that the size of the flow demanded by reluctant entrants could be very large—much larger than is politically tolerable. However, in an international regime based on textbook trading, the only efficient way to bring new countries within the regime is to impose a cap—the problem of equilibrium flows is unavoidable.

Third, the Kyoto variant of a textbook trading system could be very unstable over time. In the ideal world, countries would allocate permanent property rights and establish a system of "buyer liability" (or "issuer liability"). The market would price the risk of non-compliance or defection, and the system would enforce itself to the extent that administrators within the countries that buy the permits ensure that this liability rule is actually implemented. (Since most of the buyers, at least initially, would be in the advanced industrialized world where the rule of law is generally strong, such a system could work relatively well.) However, Kyoto's architects did not establish this theoretically ideal system. Kyoto's architects eschewed permanent allocations, setting the emission allocations for relatively short (five-year) budget periods. That approach made sense, in part because it would be extremely difficult to gain agreement on what, in essence, would be a permanent allocation of property rights. By making an allocation for only a brief time, negotiators could avoid the deadlock of attempting to find a rule that could govern permanent allocation and be acceptable to all nations.

This short-term approach, however, while the only one feasible, means that the system is regularly exposed to the risk that important parties will find their allocation inconvenient and withdraw, which in turn could ripple through the trading system as other nations follow. Evidence that this is likely to be a big problem comes from the US experience today. The main US objection to Kyoto is that it imposes an emissions cap that is particularly onerous because US emissions have risen more rapidly in recent years than had been expected, principally because the US economy has grown more rapidly in the late 1990s than most experts thought was feasible at the time when the Kyoto caps were negotiated (1997). Today, the US finds itself with inconvenient targets and no way to comply except through the purchase of large quantities of "hot air" (which is politically infeasible); the only alternative is to reject the Kyoto targets altogether. Every five years—when the allocations are negotiated—there will be fresh opportunities for defection. This time the US has been the key defector; next time it might be the EU or China. In a system that sets caps on the quantity of emissions, it is extremely hard to assure that economic and technological conditions will not change in ways that make a deal that looked acceptable when negotiated look much less appealing does not turn sour by the time of ratification or implementation. If even one party makes a wrong forecast about its future emissions—and thus accepts an allocation that it cannot implement—its defection from the regime changes the circumstances for all other nations and also changes the environmental effectiveness of the agreement.

All three of these problems stem, fundamentally, from the difficulty of reaching an acceptable allocation when the participants are extremely uncertain about future costs and have widely different preferences in their willingness to pay for reductions in greenhouse gases. To some degree, this problem is fundamental to the nature of the global warming problem. To a larger degree, the problem is a by-product of the fact that a textbook cap-and-trade system regulates the *quantity* of allowable emissions while allowing the *price* to float (potentially to very high levels); yet it is the price—not the quantity cap—that is a much better indicator of a country's real effort toward controlling emissions. Regardless of the emissions quantity, if the price to control emissions rises too high, the country is likely to defect from the Protocol.

Rethinking the Cap-and-Trade Architecture

In its current form, the Kyoto Protocol proposes a system that is unlikely to be stable and effective over the long term and yet, following the agreements in 2001, is likely to enter into force. Moreover, it may be promoting a false sense of security that enough is being done to meet the challenge of climate change. It is vital to step back and reassess the architecture of the Kyoto enterprise. Since scrapping Kyoto and starting again is an unviable proposition, negotiating a superior post-Kyoto framework that addresses the problems—emissions caps, emissions trading and allocation, and cost uncertainties—is crucial. A post-Kyoto framework could also address the need to bring in the South by ensuring effective measures that speak to their concern about being locked into a system that could prove costly and ineffective.

Broadly speaking, there are two options for the architecture of an international agreement to limit the emissions that cause global warming. The first is to set targets for emissions and leave the market to determine the cost of achieving these targets—this leads to the "cap-and-trade" architecture embodied in the Kyoto agreement. Second, governments can set the price, through tax measures, and let the market produce reductions. Other options also exist, such as direct regulation ("command-and-control"), but these are likely to be very inefficient, especially as efforts to control emissions are tightened.

Another option, a coordinated international carbon tax, advocated by Richard Cooper and others, has some strong advantages (Cooper, 1998). It would provide incentives through the market to reduce in an efficient manner the emission of carbon dioxide. The tax approach could ease the political task of allocating commitments because countries would set their obligations at exactly the level they are willing to pay rather than exposing themselves to the uncertain link between caps on emission quantities (as in Kyoto) and the ultimate cost (price) of meeting that cap. A tax by itself would not result in the huge financial flows that result when a pure trading system equilibrates.

New taxation schemes, however, are problematic and draw significant criticism. As in other international taxation schemes proposed in the past (e.g., international taxes levied on foreign currency trading such as the Tobin Tax), monitoring or oversight bodies must be put in place, agreement on equitable national implementation regulations is needed, and allocation of the tax proceeds must be agreed upon. To date, despite copious studies and debate outlining the many positive features of a carbon tax, the difficulties inherent in such a taxation scheme have all but buried the idea. Ironically, taxes make the cost of action *intolerably* transparent. The failure of a relatively tiny "BTU tax," proposed in the US under the democratic government of President Clinton perhaps provides a lesson about what the American public will accept. The greatest opponent to an international currency exchange tax has also been the US. It thus seems that a global emissions tax is a good but unworkable option, and not one that could entice the US into a post-Kyoto agreement.

A Preferable Approach

Neither of the main options appears to be feasible on its own. Textbook trading fails on the problem of allocation discussed earlier, and taxation falters on monitoring and enforcement. Instead, a hybrid of the two approaches is preferable. Governments would set targets for emission quantities, as in a cap-and-trade system; but, governments would also agree to install a "safety valve" that would ensure that the price of emission credits does not rise above an agreed level (McKibben and Wilcoxen, 1997; Pizer, 1999; Victor, 2001). The "safety valve" would work as follows. If the trading price for emissions credits goes above the agreed price, firms could purchase new permits at the agreed price from government. In the event the market price goes below the agreed price, then firms could purchase permits from the open market. In effect, the price of emission permits would be capped, making it much easier for firms and governments to anticipate the cost of compliance.

This safety valve approach would have economic advantages over the textbook cap-and-trade system because economies would not be exposed to sharp increases in price if the cost of meeting targets turns out to be higher than expected. However, the chief advantage of the safety valve is political: compliance costs become more predictable, and it is therefore much easier to allocate emission targets. When targets can be allocated along lines of marginal cost, it is easier to reduce the international capital flows that occur when a trading system is initiated. In addition, with greater cost certainty it would be easier to allocate "headroom" to reluctant participants (e.g., developing countries) that more precisely matches their need for compensation and is not loaded with liberal quantities of "hot air" as a hedge against uncertainty.

The safety valve is not high on the political agenda in Canada, but it should be. This approach would make it easy for Canada to implement a climate policy that would impose only a low (but not zero) cost on the Canadian economy. At present, the only option under consideration in Canada that would achieve this goal—low-cost implementation—involves a combination of accounting games with credits for clean energy exports (which other countries are unlikely to approve) and opening up the Canadian market to a flood of cheap Russian and Ukrainian permits (which the public is unlikely to accept when they see that it merely sends money to former Soviet states in exchange for no reduction in emissions). In contrast, the safety-valve approach would allow the government to declare a price of C$10—the most likely price for carbon credits, according to the discussion paper (Canada, 2002)—while guaranteeing that worst-case high-cost scenarios will not come to pass. The government would be able to justify this approach by pointing to two positions that are already central to the deliberations about Kyoto. First, the safety-valve approach will not undercut the substantial opportunities for low-cost emission reductions; indeed, it will focus firms' and households' efforts on finding those low-cost options while ensuring that if those options are more costly than claimed—as the skeptics argue—that the economy will not be saddled with bearing the higher cost. Second, the US withdrawal from Kyoto demands that Canada be vigilant about the impact of its policy on competition, and that requires special attention to cost. The safety valve allows the government to control cost, whereas a textbook trading system offers no limits (other than defection or accounting games) on cost.

The safety-valve proposal would be illegal under Kyoto because it might result in Canada's exceeding its emission targets. Therefore, at the same time, the Canadian government would have to announce that it will trade on the open world market (with the Russians and with developing countries through the CDM) as necessary to acquire the additional permits needed to cover any shortfall in Canada caused by the safety valve. When Canada ratified the agreement, it could have lodged a statement saying that it reserves the right to withdraw in case of credible evidence of manipulation of world permit prices. The Protocol and UNFCCC do not allow "reservations" to be lodged by parties, but price manipulation would easily fall within the purview of extenuating circumstances that a country can use to justify exiting a treaty—and it would also put permit traders on notice that collusion would be self-defeating.

As noted, Canada might also elaborate this proposal in ways that help to engage developing countries more fully—thus paving the way for a more effective climate regime in the future. For example, it could earmark some or all of the revenues from the safety valve fund (the C$10 per tonne for each new permit issued, if any) for a CDM fund. This approach would build on the successful concept of a CDM fund pioneered by the Dutch government and distributed through auction.

This approach is both elegant and powerful. It would protect Canada against high-cost scenarios that have caused so much political difficulty, especially in Alberta. It would also protect Canada against harmful trade effects with the US. The approach would also sustain the Kyoto Protocol by assuring that there is a positive price on carbon and not merely shell game trading with Russia. However, the main effect would be on the international political scene: only through actual demonstration can the idea of a safety valve gain the prominence that will be needed for it to be incorporated in the second budget period negotiations.

Getting There

The tendency in international diplomacy is to build on the status quo—to adjust, at the margin—without reorganizing the fundamentals. This tendency is particularly strong in multilateral agreements because so many parties and interests require reorientation for the collective to change course. It can be difficult for a relatively small party, such as Canada, to affect that course—especially when the major states proclaim that they are either completely in favour of the existing regime (i.e., the EU), completely opposed and disinterested in reform (i.e., the US), or muddling through with the hope that problems disappear (i.e., Japan). Added to all these difficulties is the fact that the countries that take the lead in refashioning Kyoto will pay a short-term political cost in being branded as spoilers—the US is already paying that cost, but there is still room for others to suffer that fate. Nonetheless, it is the Canadian experience that, working intelligently with like-minded countries and NGOs, it is possible to broker consensus on major issues that appear intractable.

The US has not proposed a viable alternative, or at least a framework that offers a path forward, although the policy announced by the Bush administration in January 2002 did include some positive elements. Notably, the administration proposed a greenhouse gas intensity index—the ratio of emissions to economic output—that in effect would index emission targets to economic growth. Such an approach, unlike the Kyoto system of setting targets for the total quantity of emissions, would reduce the risk that a country would find itself in the position that it could not comply with its targets because its economy had performed unexpectedly well. However, the Bush administration's plan included many disappointing elements: it ignored the need for long-term goals and visions, such as some way of working with the rest of the world; moreover, the administration undercut its credibility by proposing targets for improving greenhouse gas emission intensities in the US that were nearly identical to the "autonomous" rate of improvement in the US economy. Total emissions would continue to rise under the Bush administration plan, and there was no vision or strategy for how emissions growth might be reversed, even over the long term. The glaring deficiencies of this plan simply redoubled support in the rest of the world for Kyoto.

We argue that Canada must not fall into the same trap of assuming that, just because Kyoto's support is nearly universal, the framework is actually workable over the long term. The Canadian government should make a prominent statement about the need for reform of the Kyoto framework so that it becomes more sensitive to cost and less vulnerable to the exit of key parties, especially when that exit causes the evaporation of the incentives for developing countries to participate. We have outlined a mechanism—the safety valve approach—that meets both these criteria; but the main point is that Canada could play a critical role in offering a strategy for reform while not actually threatening the viability of the first budget period of the Kyoto Protocol (2008–2012). Successful reform will require the participation of the European Union and Japan, and neither has yet been willing to entertain reforms that would disrupt the entry-into-force of Kyoto.

The great tragedy of the unwavering support for the Kyoto Protocol as now negotiated is that it will have so little impact on global warming. Canada can have the largest impact on building an effective regime not by calling attention to that seemingly unmentionable reality, but by presenting an alternative that applies beyond 2012, a period that is much less threatening than the 2008–2012 period. If Canada does this, it may well find a receptive audience. The EU and Japan—not only the US—are aware of Kyoto's flaws, but until they have lodged their instruments of ratification, those flaws have been unspeakable.

Notes

1 Indeed, for the Kyoto emission trading system to operate, Russia must be a member. The Protocol includes a provision that allows it to enter into force only if ratified by countries accounting for at least 55% of the reported 1990 emissions of CO_2 from burning fossil fuels. Without ratification by the US, Russia's ratification is pivotal. The Russians know this and have delayed their ratification pending resolution of sundry other issues—notably foreign debts with EU members. At the time of writing (Spring 2004) there are still conflicting signals coming out of Moscow.

2 For example, in its Spring 2002 Discussion Paper (discussed below), the Federal Government makes a direct link between ratifying Kyoto and obtaining credit for clean energy exports. These exports were calculated to "displace" about 70 Mt of GHG emissions, or nearly one-third of the 240 Mt gap.

3 Canada is unlikely to convince the EU to support changes in accounting rules for two other reasons. First, the main beneficiary of Canada's energy exports is the US, which will not be a member of the Kyoto treaty. Second, Canada's rat-

ification is not essential to Kyoto's entry into force—unlike Russia's ratification, which is pivotal—thus Canada lacks bargaining leverage.

4 This figure, $100 billion, is derived by calculating the estimated number of credits per year from Russia and Ukraine multiplied by five years (2008–2012) and multiplied by the estimated price per metric tonne of carbon ($50 per ton C). The earlier number that we cited, $2 trillion, is the value of the allocation to all 38 countries computed as an asset, not the annual (or the five-year) value. For more on the calculations see Victor (2001) and Victor et al., (2000).

References

Canada. (2002). *A Discussion Paper on Canada's Contribution to Addressing Climate Change.* <www.climatechange.gc.ca/english/actions/what_are/canadas contribution/Report051402/englishbook.pdf>.

Cooper, Richard N. (1998). Toward a Real Global Warming Treaty: The Case for a Carbon Tax. *Foreign Affairs*, March/April.

Energy Information Administration. (2002). *International Energy Outlook 2002.* <www.eia.doe.gov/oiaf/ieo/>.

McKibben, W.J., M. Ross, R. Shackleton, and P.J. Wilcoxen. (1999). Emissions Trading, Capital Flows and the Kyoto Protocol. *Brookings Discussion Papers in International Economics* 144 (February).

McKibben, W.J., and P.J. Wilcoxen. (1997). *A Better Way to Slow Global Climate Change.* Brookings Policy Brief No. 17. Washington, DC: Brookings Institution.

Office of the Press Secretary. (2001). Text of a Letter from the President to Senators Hagel, Helms, Craig, and Roberts. <www.whitehouse.gov/news/releases/2001/03/20010314.html>.

Pizer, William A. (1999). *Choosing Prices or Quantity Controls for Greenhouse Gases.* Climate Change Brief No. 17. Washington, DC: Resources for the Future.

Victor, David G. (2001). *The Collapse of the Kyoto Protocol and the Struggle to Slow Global Warming.* Princeton, NJ: Princeton University Press.

Victor, David G., Nebojsa Nakicenovic, and Nadejda Victor. (2001). The Kyoto Protocol Emission Allocations: Windfall Surpluses for Russia and Ukraine. *Climatic Change* 49: 263–77.

What Can Individuals Do?

Harold Coward

Among the world's developed countries, Canada has the highest levels of consumption per capita, and it will have difficulty meeting its Kyoto Protocol target. As Weaver points out in chapter 2, lifestyle changes are needed if Canada and the world are to reduce carbon dioxide (CO_2) emissions and thus allow the Earth system to equilibrate with higher levels of greenhouse gases and warmer overall temperatures. Lifestyle changes are also required from Canadians if global or intergenerational equity (justice) is to be achieved. As Lonergan argues in chapter 3, Canada needs to reduce its CO_2 output and increase its carbon sinks to meet its Kyoto target, but also to do justice to future generations and to people in developing countries. Equity requires that Canadians reduce their per capita emissions so as to allow room for developing countries to expand theirs until there is what Lonergan calls "global convergence." Although some of this reduction in CO_2 emissions can be obtained by "technological mitigation" (as McLean shows in chap. 6), lifestyle change will still be required from Canadians if intergenerational and global equity is to be reached. This chapter explores various ways in which Canadians can change their current heavy consumption lifestyle patterns. We will first look at an ethical framework for knowing "what ought to be done" and then explore resources in religious traditions for doing what is right and good.

An Ethical Framework for Knowing What Ought to Be Done

Most Canadians think of themselves as following an ethical approach to life. What response from us to the challenge of climate change would be ethically right? One ethical argument for lifestyle change can be made by examining the consequences of our current consumption patterns for people living in our own family, city, or country; for people in other countries (e.g., especially in developing countries); for future generations (e.g., our children and grandchildren); and for the environment valued for itself (e.g., earth, air, water, plants, and animals having value along with humans). For Canadians as individuals, corporations, or governments, the decisions we make as we attempt to deal with the challenge of global climate change (chap. 2), its human implications (chap. 3), and the economic factors involved (chap. 7) can be either ethically right or ethically wrong (Hurka, 1993, p. 23).

Hurka points out that ethics need to be distinguished from opinion. Surveys to determine what people think is right or wrong about climate change, for example, describe opinions rather than ethics. Too often governments make decisions based upon polls of people's opinions rather than a careful study of the ethical issues involved. Ethics is about values apart from people's opinions. Ethics assumes that some beliefs about right and wrong may be incorrect, and the study of ethics attempts to discover which ones are correct. In short, there is right and wrong above what people *think* is right and wrong—beyond people's opinion.

Ethical decisions require that we combine the scientific, social, and economic facts relating to the threat of global climate change with general ethical principles that indicate right and wrong in all areas and thus lead to specific policy recommendations. We can, of course, argue over which ethical principles should be employed in such an analysis, and the employment of different principles could lead to different ethical conclusions and different policy recommendations. This difficulty can be dealt with by selecting ethical principles which are not radical or speculative but are widely accepted. In this way, the policy proposals developed by an ethical analysis can be convincing to most people. Using "an analysis of consequences of actions" as an approach allows us to move from areas of least controversy and broad agreement (e.g., impact on one's own family and country) to areas where the policy conclusions are more radical and the agreement less general (e.g., impact on the environment). General policy decisions in response to the challenge of climate change can favour either adaptation, or avoidance. With adaptation, we keep burning fossil fuels, let global temperatures rise, and make whatever changes this requires: move people from environmentally damaged areas, build sea walls, and so on. With avoidance or mitigation, we make every effort to stop warming from occurring, by reducing our use of fossil fuels, by using mitigating technology (chap. 6), and by making lifestyle changes. As we shall see, ethical responses to climate change strongly favour avoidance over adaptation in individual, industrial, or government decision making. However, it is unlikely that either pure strategy is possible. According to current estimates (see chap. 2), pure adaptation would result in a temperature and sea-level rise that would be faster than any in the last ten thousand years, and would be devastating for many human and non-human communities; but pure avoidance—reducing warming to zero—would be enormously expensive or even (with population growth) impossible to achieve. Therefore, an ethically acceptable goal will likely involve some mixture of adaptation and avoidance or mitigation.

Adopting the ethical principle of considering the consequences of our actions means that if an act or policy has good consequences, then this counts ethically in its favour; if it has bad or disastrous consequences, this counts ethically against it (Hurka, 1993, p. 24). How does one decide which consequences are good? One popular principle from utilitarian ethical theory says that good decisions are those that maximize the best consequences so as to produce the greatest good possible. Other philosophers (e.g., Rawls, 1971) care not only about the total good a choice or policy will produce, but also about the breadth and equality of its distribution. A less demanding "satisfying principle" (from the idea of "making satisfactory") gives each of us "the duty only to bring about consequences that are reasonably good, either because these consequences are above an absolute threshold of satisfactoriness or because they represent a reasonable proportion of the most good the agent can produce" (Hurka, 1993, p. 25).

How do these ethical principles about "consequences" apply to decisions about climate change? Where actions such as burning fossil-fuel and generating CO_2 foster global warming, with its negative consequences such as sea-level rise displacing billions of people (see chap. 2), it is clear that our ethical duty is to avoid such a result. Hurka summarizes the situation:

> If an act or policy involves some risk of bad consequences this is a reason to avoid it, and this reason is weightier the worse the consequences are and the higher the risk. If the consequences are extremely bad, even a small risk of producing them is a reason to avoid the act and to accept some costs in doing so.... If the result of allowing climate change would be disastrous, it is prudent to avoid this result even if we are not certain that it would come about. (Hurka, 1993, p. 25)

In simple language, "better safe than sorry" when the potential consequences of climate change are so serious. Now that we have a clear idea of how ethical judgments can be made by examining the consequences of our actions or policies, let us return to the questions relating to lifestyle change with which we began—beginning with our immediate family, and then widening our concern to include others elsewhere in the world, future generations, and finally nature itself.

Consequences: Humans here and now

Of course we will all begin by caring about how climate change will affect each of us, our families, and our homes. This is simply our own self-interest and does not really count as *ethical*. Our behaviour becomes ethical when we make decisions regarding climate change that will benefit and not harm others living in our neighbourhood, city, and country. What counts as benefits or harms? These clearly depend on more than one's income. Money is only a means to other things we value. Secular philosophy gives at least two answers. As Hurka puts it, "According to *welfarism*, humans are benefited by whatever gives them pleasure, fulfils their desires, or contributes to something describable as their 'happiness.' *Perfectionism*, by contrast, equates the human good with knowledge, achievement, love, virtue and other states that it values apart from any connection to happiness" (Hurka, 1993, p. 25). It is how we develop our potential, not how happy we are, that ultimately matters. When we focus only on the present and the effects of our actions on our own families, cities, and country of Canada, many of the most harmful results of global climate change seem not to count—for example, damage to the environment from a rise in global temperatures (killing organisms and ecosystems) does not matter according to this principle since only humans and not nature have ethical standing. Further, since the most severe consequences of climate change may affect future generations, such harm is ignored by the "humans here and now" principle. However, this principle does not capture some of the sacrifices this generation would have to make if, for example, reducing the use of fossil fuels to avoid CO_2 production should result in economic costs. Ethical analysis of the "humans here and now" principle tends to favour adaptation rather than avoidance or mitigation behaviour—it does not foster lifestyle change and simply sits still while climate change continues. It would, however, support technological mitigation measures (see chap. 6), such as increasing the efficiency of heating, lighting, cars, and electrical generation plants—as long as they did not cost too much. To reach an ethical approach that would argue for less adaptation and greater lifestyle change, we must extend our concern for consequences out beyond "humans here and now" to the wider principle of "humans everywhere in the present."

Consequences: Humans everywhere in the present

This principle suggests that to maximize the good and be egalitarian in our ethics, we must be as concerned about the benefits and harms wrought by climate change in other countries as we are about those in our own. Extending the analysis to other countries strengthens some arguments for avoidance and mitigation. It suggests that, as countries like China and India industrialize, we in Canada should help them by providing energy-efficient technology (at a cost they can afford). To be egalitarian about sharing the benefits of electricity and a higher standard of living means that we will likely have to alter our lifestyle and pay more for everything. To achieve this global benefit will cost developed countries like Canada more; but the result will be an increase in the standard of living for people in developing countries—an ethical result. Some sacrifice will be required from Canada and other developed countries to meet the goal of enabling developing countries to industrialize and achieve a higher quality of life with energy efficiency so that additional CO_2 production and damage to the environment is minimized. The ethical challenge is to balance competing claims for equality among nations.

Consequences: Future generations

It is when we think of the effects of climate change on future generations in Canada and elsewhere in the world that the need for mitigation and lifestyle change is strongest. The predicted rises in sea level, the destruction of traditional habitats and industries, and the loss of biodiversity push the ethically acceptable climate policy strongly towards mitigation rather than adaptation. We want to pass on a healthy environment and a sustainable lifestyle to our children and grandchildren. Just as an egalitarian ethical principle argues for equity between nations, so also we must ensure that there will be equity for future peoples—"seven generations into the future" to quote one Aboriginal teaching. Other Aboriginals argue that equity extends from three generations before us (great-grandparents) to three generations after us (great-grandchildren).When we factor in worries over population growth, which threatens to increase our current global population in 50 years from 6 to 10 billion people, the ethical challenge becomes very demanding. While we need to cut back our consumption now to create room for developing countries to industrialize, we also need to restrain ourselves even more severely if we are to create a lifestyle that is sustainable for large population increases in the future.

Consequences: Nature valued for itself

We have widened our application of ethical principles from our own families and country to people everywhere and to future generations—but what about the environment, nature itself? In the recent past, our assumption has been that changes to the environment mattered ethically only if human life was thereby affected. Even the 1987 Brundtland Commission report, which championed sustainability, boldly asserted that the well-being of people is the ultimate goal of all environment and development policies (World Commission on Environment and Development, 1987, p. xiv). A more radical view argues that we need to care for the natural world not just as a means to better human lives but as an end in itself. When we adopt the ethical position that nature has intrinsic value, the main problems to be dealt with are of two kinds: (1) overpopulation by humans, which threatens to squeeze out other species and overwhelm the carrying capacity of the earth; and (2) the rapid rate of climate change (faster in the last few decades than in previous history; see chap. 2), which threatens many forms of life that require slower warming to be able to adapt successfully. Thus we are in danger of losing both individual species and whole ecosystems: "This will be bad both on an individualist environmental view—where individual animals will suffer or find their natural life-activities impossible—and on a holistic view, where complex and fragile ecosystems, such as in the Canadian Arctic will disappear" (Hurka, 1993, p. 32). Concern for nature valued for itself leads us even more strongly to embrace the approach of avoidance or mitigation in our production of pollutants that foster climate change. This ethical approach requires even more human sacrifice if ecosystems such as the Arctic or the Southern Prairies are to be preserved (see chap. 4).

The ethical principles supporting the valuing of nature for itself can take several forms. The Welfarist approach (mentioned above) argues that the pain and pleasure of animals has at least equal importance to the pain and pleasure of humans. Thus climate changes which cause suffering to animals are to be avoided as are changes bringing misery to humans. A Perfectionist ethics gives value to a wider range of lifeforms than does Welfarism, and at the same time allows more distinctions to be made among them. For example, while valuing the flourishing of insects, fish, and mammals, the perfectionist can argue that their value is less than that of humans because of the human'shigher mental and rational capacities. A third approach, which is referred to as holistic environmental ethics, was given its classic statement by Aldo Leopold: "A thing is right when it tends to preserve the integrity, stability, and beauty of the biotic community. It is wrong when it tends otherwise" (1970, p. 262). This approach takes the bearer of intrinsic value to be the whole of nature, of which humans are simply a part—e.g., ecosystems such as the fish in their ocean habitat in relation to the fishing communities and the climate that sustains them all. Such ecosystems can extend out to include the entire earth biosphere and, according to holistic ethics, grant humans (either as individuals or groups) ethical significance only as contributing to the harmonious working of the overall whole. In this view, if humans by their behaviour both overpopulate and overconsume, they may be in danger of being wiped out as a species to save the functioning of the earth's ecosystem. In the holistic view, ethical standing belongs not just to individual organisms or species but to the interrelated ecosystem wholes that they compose. As we shall see in the next section, the holistic secular approach has much in common with Eastern religions and the Aboriginal traditions.

Resources in Religious Traditions for Doing What Is Right and Good

Another source of guidance for making lifestyle choices in relation to climate change are religious and spiritual values. Although more Canadians may be swayed by secular ethics, for well over half the world's population traditional religions still construct the worldview that guides how people deal with questions of population growth (reproduction), consumption, and the environment—as we have seen, all crucial factors in decision making with regard to climate change. There are additional reasons for including religious and spiritual values. First, just as philosophy, law, and economics can offer guidance based upon a long history of human thought, so also religion has been and remains a major part of human civilization; therefore, its wisdom should be considered. Second, some states are religious rather than secular (e.g., Iran and Pakistan); therefore, knowledge of religious responsibility in relation to nature, consumption, population growth, and other factors, can help in appealing to such states for international cooperation on climate change problems. In the following analysis of religious traditions, we will focus on their teachings about what individuals can or should do to make a difference in response to the challenge of climate change. Beyond Hurka's ethical principles, what are the added teachings offered by religions to convince their members that we are really interconnected with other humans and all of nature, and therefore we ought to take responsibility for the impact of our actions on other humans (living now and in the future) as well as on the natural environment.

Aboriginal spiritual traditions

Studying religious and spiritual traditions highlights the radically different ways we human beings see ourselves in relation to the world around us. When modern technological societies confront aboriginal groups with a desire to "develop" land that has been their home, a clash of perspectives occurs. The "developer" sees the non-human world as passive, unexploited, and in need of human initiative for its potential to be realized. The implied ethic in the developer's view is that it is the duty of humans to improve nature by taming the wilderness that has held previous generations in terror and bondage. With our more powerful technology, humans today can do what earlier generations sought to do, namely, subdue and master nature. Aboriginal cultures, while often very different from one another, share a quite different approach. Rather than desiring domination, their spiritual teaching is that humans need to live in harmony with the land and its non-human beings. Aboriginal traditions aim at finding one's place in nature by knowing the story of the land or sea, its animals, and the complexities of its seasonal rhythms so as to live in ways that do not disrupt its divinely given harmony. Although this knowledge is used in practical ways, such as in successful hunting, the desire to dominate the land is not a strong theme. Rather, Aboriginal teachings tell that nature is alive and full of non-human species that have formed the land, air, and water, and continue to permeate it. For humans to live in nature ethically, it is necessary to see themselves as part of nature and to relate properly to the other spirit beings that form nature: "Not only does everything have spirit, in the last analysis all things are related together as members of one universal family, born of the father, the sky, the Great Spirit, and one mother the Earth herself" (Callicott, 1989, p. 186). Thus we must treat all of nature as we would treat the members of our own family and recognize that there is a spiritual aspect to all natural things. The ethnical responsibilities due to our own family or tribe are extended to include one's "natural relatives," which make up the environment.

This ethic allows the eating of animals and plants needed for survival if it is done with sincere respect and kinship feeling, but rejects the abuse of these other species for personal overindulgence or selfishness. This same ethic extends itself to our focus on the atmosphere. The air makes possible our survival as humans and is necessary for the well-being of other members of our cosmic family, the animals, plants, and oceans. Pollution of the air, caused by consuming more than is essential for life, is unacceptable in the light of our responsibilities to other humans and to the animals, plants, and oceans that will be affected both now and in the future. Aboriginal traditions would insist on extending Hurka's framework of ethical responsibilities from humans here and everywhere to all aspects of nature "seven generations into the future," to use an Aboriginal rule of thumb. Adaptation to global warming would be acceptable to the extent that the warming is the result of ethical human behaviour necessary for survival; but when warming is the result of human behaviour motivated by the selfish desire for more food and material possessions than necessary for survival, avoidance and mitigation are preferred. Production of children should be limited by consideration of what the earth can support without damage to the other species (Sewid-Smith, 1995, p. 67).

Asian religious traditions

Hinduism

Like Aboriginal traditions, Hinduism sees the natural world around us as an environment alive with spirits to whom reverence is appropriate. This is seen in the earliest Hindu scriptures, the Vedas, and continues till today in the many varieties of Hindu belief and practice. There is a strong tendency for Hindus to regard all of nature as a sacred living being. One key Hindu scripture, the Bhagavad Gita, describes the world as God's body. Thus the land, its rivers, and its atmosphere, together with the humans, plants, and animals that dwell in it, are parts of God's body. As humans, our responsibility, say Hindu law books, is to keep the cosmos (God's body) pure and orderly. For example, the pollution of lakes and rivers is prohibited. Human intervention into nature is seen as lawful only when it does not disrupt the cosmic order (*rita*). Nature, as God's body, is carefully protected from human exploitation. Through its scriptures and law codes, Hinduism views the universe as a series of interrelated systems in dynamic equilibrium, within which humans must play their part as responsible participants (Crawford, 1989, p. 30).

This ideal of restraint leading to renunciation has functioned in India for centuries. As a result, Indian culture has a deeply rooted "conservationist ethic." Hindu ethics do not reject technology or material possessions, but see them as having a restrained but proper place in the cosmic order of God's body. One could perhaps look at Mahatma Gandhi as an embodiment of the Hindu ideal of restraint. He attempted to guide India between the extremes of no growth at all and growth for material values only, choosing instead a course of selective growth guided by spiritual goals. His Hindu Vaisnava background—with strong Jaina influence—led him to advocate social models that balanced economic and environmental needs.

In terms of the range of duties listed in Hurka's ethical analysis, it is clear that Hinduism views us as having duties not only to humans, of whatever time and place, but also to animals, plants, and elements of the environment, all of which are taken to be God's body and therefore to have intrinsic value. Since the atmosphere is also seen as a valuable part of God's body, pollution of it in ways that lead to global warming is not acceptable to Hindu ethics. Avoidance or mitigation rather than adaptation is clearly the counsel of Hindu teachings. This counsel presents a serious challenge to India's modern cities, such as Bombay, where carbon dioxide pollution runs rampant. In spite of its fine ecological teachings, India, like the West, has ignored these teachings in its rush to modernization; it has not followed Gandhi's ideal of restraint.

Buddhism

Perhaps more than any other religion, Buddhism focuses upon an analysis of the individual and what an individual can do to live an ecologically responsible life in response to the challenges of climate change. Throughout the many varieties of Buddhism two themes dominate—suffering and non-violence—both of which are important in regard to climate change. In the Buddha's analysis of life, "suffering" (*dukkha*, which includes both physical pain and mental frustration) is our most common and pervasive experience. The major cause of this suffering is our desiring, which is insatiable. No matter how many material possessions or how much pleasure we experience, we are always left wanting more. One car per family is good, two is better—and we still might want a mountain-bike, a boat, the latest skis... and so it goes in a never-ending list of wants. All of this "desiring" is driven by a false notion of ourselves as a separate ego, "I," or "me," that wants the possessions and power to make life turn out the way we want it to be; but no matter how hard we try, we find we cannot control this cosmos, which is in a state of constant change. The Buddha's enlightenment experience showed him that by giving up the notion of a permanent ego, "I" or "me," and the insatiable desires for more that it produced, the experience of life as suffering was also overcome. Then life, this ordinary everyday life, is seen to be what it really is—beautiful, compassionate, and harmonious (*nirvana:* this life minus ego selfishness).

How does one achieve this "*nirvana*-vision" of life that changes it from suffering to beauty and compassion? By the practice of meditation and following the ethic of non-violence (*ahimsa*). This is where the relevance of the Buddha's teaching for the challenge of climate change becomes clear. Through meditation, the Buddha came to the realization that he was not a separate ego, but an interdependent part of the constantly changing cosmos. To hurt another part—be it a person, animal, plant, or the atmosphere—to satisfy our selfish desires is tantamount to hurting ourselves. So meditation teaches us the ethic of non-violence to the rest of the cosmos, of which we are but interdependent parts (Cook, 1989). To keep the universe and our world functioning in its divinely ordered beauty, harmony, and compassion, it is our individual responsibility to use our free choice in non-selfish and non-violent ways. Through such choices, we would stop inflicting pain on other humans and animals. We would protect the forests, the soil, the water, and the atmosphere. In so doing we would be in *nirvana*, which is not some other world, but this world minus ego-selfishness. In terms of Hurka's ethical framework, Buddhism clearly establishes duties that extend to include humans and animals everywhere, now and in the future. Plants and the rest of the environment are valued only slightly less than humans or animals but are seen to be crucial interdependent parts of the cosmos and thus essential to the harmony of the whole. Regarding global warming, the counsel of the Buddha would require that we as individuals practice avoidance or mitigation rather than adaptation. As one Buddhist scholar summarizes, this means we each must negate our selfish desires for more children or material possessions when having them would harm other persons or the environment (Gross, 1997).

Chinese religions

Like Buddhism, Chinese religions view the cosmos as a harmonious process that exhibits three basic motifs: continuity, wholeness, and dynamism (Wei-ming, 1989, p. 69). Rather than thinking of ourselves as autonomous individuals, we are taught to see ourselves as part of an extended family that includes ancestors, and future generations, and extends outward to embrace all of nature in its changing seasons. Within this vision, all things are part of the same whole—the interrelated and interdependent cosmos. Thus our choices and behaviour are to focus not on our wishes as separate individuals but upon keeping ourselves in harmony with the wholeness (the *Tao*) of the cosmos, which is pervaded by vital energy and is in constant change. The pattern of this change is best seen in the rhythm of nature and the changing seasons—a pattern we should "flow-with," rather than attempting through our individual action to go "against the flow." This natural "flow" or pattern is often referred to as the "course of heaven" (*t'ien-hsing*). It is vigorous in nature and is always tending toward balance and harmony—as seen in the idea of the transformation of *yin* and *yang*, the two polarities that in Chinese thought compose nature. *Yang* includes the male, the sun, fire, heat, heaven, creation, dominance, spring, and summer. *Yin* stands for the female, the moon, cold, water, earth, nourishing, recessiveness, summer, and winter (Chan, 1969). All of this is symbolized in the familiar *yin-yang* diagram of a disk on which *yin* is represented by the dark side and *yang* by the light side. The S-shaped line that divides the two halves suggests constant motion, the idea of the primal forces of nature in their constant interaction. *Yin* and *yang* are complementary rather than being in opposition—they are constantly being transformed into one another in a rhythmic and harmonious fashion. The challenge for us as individuals is to decide to go with the flow of nature symbolized by *yin/yang* rather than attempting to go against it by following our selfish individual desires. The human forcing of climate change outlined in chapters 2 and 3 is, in the Chinese view, a result of our attempting to go against the flow of nature rather than "tuning into nature" and "going with the flow." In terms of Hurka's ethical framework, the approach of Chinese religions requires sensitivity to humans, animals, plants, and all of nature (including the atmosphere) in making ethical choices.

The Abrahamic religions: Judaism, Christianity, and Islam

Judaism

In Judaism, humans occupy the ambiguous position of both being a part of creation and exercising dominion over it. As Genesis 2:15 puts it, "The Lord God took the man and put him in the garden of Eden to till it and keep it." Humans are not only part of nature; God has given them the role of being partners in creation. Judaism recognized the dangers this posed for the selfish exploitation of nature by humans: "The rabbis... were not unaware of potential conflicts over 'ownership,' seeing the natural tendency of people to forget the greater unity that they share with creation and begin to act as lords themselves, exploiting the earth for short-term gain while sacrificing life in the process" (Shapiro, 1989, p. 180). Consequently, blessings were required along with special offerings before humans could use the fruits of creation. Such blessings recall to mind God's ownership of creation and caution against the misuse of nature.

In the Bible, God reminds humans of God's ultimate ownership of the land: "The land shall not be sold in perpetuity, for the land is mine; with me you are but aliens and tenants" (Leviticus 25:23). The Bible views humans as part of nature—a part that has self-consciousness. The danger is that human self-consciousness sins by thinking only of itself and forgetting its humble place in God's larger scheme of creation. Selfish use of the land can lead to its infertility (Isaiah 24:4–5), and safeguards are built in to protect the land. Exodus 23:10–12 requires the land to lie fallow every 7th year and Leviticus 25 every 50th year as well. Indeed Leviticus seems to suggest a 50-year cycle where all hierarchy is abolished and everything renews itself on the basis of harmony between God, humans, and all of nature. All begin again from a position of peace. Such a vision is particularly present in prophets such as Isaiah and Ezekiel.

Apart from such scriptural visions, Judaism seems often to have been dominated by a practical, self-interested approach to nature (Segal, 1998, pp. 4, 8). Deuteronomy 10:10 commands Israelite armies when attacking a Canaanite town not to destroy its fruit trees by wielding an axe against them. Although the soldiers may take food, they must not cut the trees down. Segal notes that the rabbis elaborated this practical rule to the prohibition of *bal tash-hit* which extends the ban on wastefulness to include other foodstuffs, clothing, fuel and water, or any other useful resource. Today's rabbis presumably would have little difficulty including pollution of the atmosphere within this ban. Jewish tradition seems quite aware of our dependence on our natural environment, and has set down concrete measures for ensuring its physical continuity as well as its quality. All this was done in the consciousness that God did indeed create in this world goodly creatures and fine trees to give pleasure to humans. The life-sustaining quality of the environment is not simply an obligation for the benefit of the current generation: "God expects us to turn over the land to the next generation, to our progeny, with all its resources intact. The land is not ours to dispose of, but only to make use of with reverence and responsibility" (Schorsch, 1992, p. 3; see also Levy, 1995, p. 96). The same principle would be applied to the atmosphere.

In the face of the pressures of proliferating population, what wisdom does Judaism offer to help sustain life on this globe? Many thinkers call upon the mystical thought of the Kabbalists in responding to this question. The answer offered is that humans must learn to limit themselves—their rate of reproduction, their use of natural resources, and their production of fouling wastes. The example to emulate is the Kabbalist vision of how God created the world. If God is everywhere, reasoned the Kabbalists, then the only way God could create would be by an act of *tsimtsum*—voluntary withdrawal or limitation to make room for creation. Similarly, we as humans must withdraw or limit both our reproduction and our wants so as to make room for coexistence with our environment in this and future generations.

If we think of our responsibility to the environment from the perspective of Judaism and in terms of the schema outlined by Hurka, it is evident that Jews see themselves as having duties to other humans presently living (anywhere in the world) and to future generations. Less clear is the degree to which humans have duties to animals, plants, the environment, and its ecosystems. Scripture and commentaries seem to agree with the practical position that humans should respect those aspects of the environment necessary to sustain life. From this perspective nature is valued mainly because of its usefulness to humans. At the opposite extreme are the Kabbalists. Whereas the Kabbalists would adopt a pure avoidance approach, other traditions within Judaism would likely argue for a blend of avoidance and adaptation in response to environmental degradation.

Islam

Muslims are directed to use their God-given intelligence to understand themselves in relation to nature and to satisfy their basic, instinctual desires while remaining in harmony with nature. Working to make the earth more fruitful or to further highlight its beauty is judged an act of worship and service to God. As such, it must be entered into with selfless dedication. When one's actions are motivated by the correct intentions, bodily and aesthetic pleasures are viewed positively as a foretaste of the paradise to come. Therefore the Qur'an commands, "Eat, drink and enjoy yourselves, but do not abuse" (Al Faruqi, 1989, p. 228).

The Islamic approach guards against any temptation to deify nature or worship any of its elements (e.g., the sun). Indeed, Islam sees nature as joining with people in the worship of the one God, their Creator. God's lordship is understood as making all of nature and all people inherently *Muslim*: "There is the concept of a natural, cosmic *Islam*, in which stars and molecules, species and elements, plants and creatures, all 'worship' by their very conformity to the laws of their being" (Cragg, 1977, p. 11). Nature not only worships God but, by its very existence, displays God's potentialities and attributes. As Sayyid Ahmad Khan, an Islamic writer of the nineteenth century, saw it, the potentiality of nature is empirical evidence of the goodness of God (McDonough, 1984, p. 35).

As was the case with Judaism, it is perhaps within mystical thought that the human relationship with nature and God is most clearly seen. Nasr (1987, p. 346) draws our attention to the Sufi idea that nature is the manifestation of God's compassionate breath. Nature in its innocence manifests God's breath through its regularity and beauty. The human who surrenders to God discovers the compassionate breath within and sees its presence all around in the creatures and entities of the cosmos. Nature is thus a vehicle by which humans can be brought to see God's truth, beauty, and compassion. Both the mystic and the scientist, through their respective disciplines, are understood by Islam as capable of seeing the divine truth inherent in nature. For Islam, the goal of both mysticism and science is to unveil the divine reality within nature and to enable humans to experience the unity of themselves and nature in the Qur'anic revelation of God.

What then is our human responsibility to nature according to Islam? The Islamic view might be stated simply as "God possesses the cosmos, humans have it on trust!" First, although nature is seen to have been created by God for the benefit of humans, it is clear that it is not to be used by humans for selfish purposes. Second, nature in itself is innocent and is a manifestation of God. Thus, nature is both a source of grace to humans and, together with the Qur'an, a revelation of God's truth. Third, nature, as well as being innocent, is also fragile. Its balance can be easily upset, especially by human wickedness. Natural disasters such as floods, hurricanes, fires, and earthquakes are interpreted by some Muslims as warnings from God that people are embarked upon a fundamentally wrong course of action, and the disasters of the greenhouse effect or fishstock failures could be similarly understood. Humans, as custodians of nature, are free to satisfy their needs only with an eye to the welfare of all of creation. In terms of Hurka's ethical criteria, Islam maintains that we have duties to all humans, present and future, and to all animals, plants, organic, and inorganic aspects of nature—for nature is an innocent manifestation of God over which we as humans are given the responsibility of obedient custodians (as defined and revealed in the Qur'an). Humans are seen as having God-given rights to use nature in satisfying their proper needs, but humans do not have the right to use nature in a way that would upset the divine balance present in creation. Global warming would be judged by Islam to be a sign of just such abuse, and a warning that God, through nature, may be about to strike back. Thus, humans had better quickly change their disobedient ways. Obedience to the will of God, in the view of Islam, requires avoidance or mitigation rather than adaptation, and clearly requires a lifestyle change from most of us.

Christianity

Christianity views both humans and nature as created by God, with nature's purpose being, at least partly, to provide for the needs of people (Psalm 105). In this, Christianity is like Judaism and Islam. Nature, by its very existence, praises God and manifests God's awesome powers (Psalm 148). Jesus saw nature as having value not in itself but only in relation to God's purpose. Human misuse of God-given freedom brought on the Fall; God's grace in Jesus Christ restores to humans the opportunity of living a righteous life in relation to nature and God (Romans 8:1–4). For the Christian, it is the grace of Christ that enables one to see nature not from the selfish perspective of fallen humanity but from the perspective of God. What is required of individual Christians in times of environmental crisis is that they see from God's perspective—"see like God." A good way to begin is by looking through the eyes of Jesus. Jesus says, "Consider the lilies, how they grow; they neither toil nor spin; yet I tell you, even Solomon in all his glory was not clothed like one of these" (Luke 12:27). The assumption behind Jesus's teaching is that God does "clothe" flowers with splendour and lovingly cares for all of creation—plants, animals, atmosphere, and all else. Our role is to be God's co-workers on earth and to use our freedom along with our scientific and technological knowledge to care for nature and in so doing bring about God's plan for creation. That is the way humans are to understand the "dominion over nature" given them by God (Genesis 1:28).

Seeing like God, then, is learning to love and care for all of creation, not merely that which involves one's personal welfare (Nash, 1991, pp. 139–61). To consider the environment only in the narrow view of its usefulness is to be greedy and self-centered. Such selfishness fails to see that all of creation is interdependent, as the prophet Hosea observes when, after describing the selfishness of Israel, he concludes that "Therefore the land mourns, and all who live in it languish; together with the wild animals and the birds of the air, even the fish of the sea are perishing" (Hosea 4:1–3). Such words seem prophetic in regard to our loss of the cod fishery on Canada's east coast, the depletion of the salmon in the West, the droughts on the prairies, and the health alerts resulting from air pollution in cities like Toronto and Montreal. If the interrelatedness and interdependence of humans and the environment are not kept in mind, then selfish action brings destruction upon nature, of which we humans are a part. However, the biblical teaching is that God acts on behalf of the whole creation, and this is the example we are to follow. For example, seeing as God sees requires that we become sensitive to the presence of God's spirit everywhere in the world upholding life. God is not an absentee landlord but is actively present in creation, "self-revealing and grace-dispensing, leaving signs and making the divine presence felt in things" (Nash, 1991, p. 111). That God cares for creation in its detail (the atmosphere, oceans and rivers, the land, individual animals, and plants) means that nothing short of similar ethical action is required from individual Christians as they respond to the challenges of climate change.

With this understanding of what is required if we "see like God," Christians might well ask how so many churches and individuals have come to view nature as simply an environment subject to human exploitation and "development." There is great diversity within Christianity today, from the conservative to the radically progressive. Within that diversity, however, many Christian thinkers are accepting responsibility for fostering much of the world's over-consumption and environmental devastation, and at the same time they are seeking to recover a mainstream Christian approach which "sees as God sees" and becomes a self-critical force for justice, peace, and the maintenance of the integrity of nature. For example, Rosemary Ruether points out that nineteenth-century European Protestantism made the Bible seem to be more anti-nature than it is by interpreting the scriptures with a dualism between human history and nature. Any divinity or spirituality associated with nature was seen to be a "false god." More recent scholarship, however, has shown that this dualism distorts the biblical perspective. Nature in the Bible is not subhuman, nor is human history set against nature (Reuther, 1992, p. 208). God's love is present when God makes the earth and the waters, fish, animals, and humans, all of which God sees as good. This strong ecological view was lost during the medieval period when the Church became more concerned with political power; but the vision was never completely lost as individuals like St. Francis of Assisi demonstrate.

Many biblical passages direct Christians to follow redemptive or restorative behaviour in their interaction with nature. The "sabbatical" and "jubilee" prescriptions in Leviticus 25 are aimed at giving nature a reprieve from human exploitation, so that it could rejuvenate, and so that human relationships could return to a just balance. The messianic vision in Isaiah 65:17–25 also offers a model of the restoration of right relationships between animals (the lion and the lamb), between people, and between people and nature (including ourselves and the atmosphere). The vision of a renewed right relationship between humans, nature, and God is also found in St. Paul's "childbirth image" in Romans 8:18–39, in which the whole of creation is seen to be "groaning" in hope towards redemption in a "new creation." The role of the individual Christian is to work with God's Spirit as it moves in the world to actualize this "new creation."

Paul says that God's Spirit intercedes on our behalf with groans that are part of the direct "experience" of God, the very same as it is experienced by the human and non-human parts of creation. The "groans" relevant to climate change arise out of our exploitation of the earth's carbon resources, the air pollution, and the global warming that have resulted (chap. 2). Nature's "bondage to decay," as Paul put it in Romans 8, includes the destructiveness of human technology in our use of carbon energy sources. This "groaning of nature" under human exploitation need not deny all human use of the earth's resources. Rather, the "groan of nature" recognizes that nature is in a dynamic equilibrium. Working as co-creators with God requires that individual Christians, in their necessary use of the earth's resources to sustain life (plants, animals, and atmosphere), do so in an ecologically respectful way. This requires redemption from our selfish desires that drive our unsustainable and disrupting decimation of nature. It also requires that we turn our technological expertise to redemptive goals (e.g., cleaning up car exhaust, developing fuel-cell energy technology, cleaning up oil spills on sea and land). Recent developments in energy and genetic engineering suggest that our new technologies may have considerable redemptive power (see chap. 6) which, together with changes in consumption and reproduction patterns, could help us to restore ecological equilibrium.

What can individual Christians do? By disciplining themselves to "see as God sees" Christians in Canada should voluntarily reduce their high levels of consumption (including energy) by simplifying their needs—here the "Voluntary Simplicity" programs popular on the West Coast offer a useful beginning (Northwest Earth Institute, 2000)—and by making use of "redemptive technology" (e.g., fuel-cell vehicles) even though it may cost more. In the face of the looming crisis of overpopulation, Christians can limit their own reproduction and, through their churches, sponsor the immigration of refugees into Canada from overpopulated and economically impoverished areas of the world. In the use of natural resources, they can be guided by the rule to leave enough for seven generations into the future. In terms of Hurka's ethical framework, the Christian approach embodies duties to humans everywhere now and in the future. Although other parts of nature are seen as having been created by God for human use, they are also valued as parts of God's good and beautiful creation. Although, through their own willful selfishness, humans have seriously damaged the divinely created balance of nature, by opening themselves to the grace of God's spirit as it works to heal the earth's ills, humans can work with God in the redemptive establishment of a "new creation" on earth. In working for this end, the response of individuals to global warming, over-consumption, and population pressure leans strongly toward practices of avoidance or mitigation, and thus requires major lifestyle changes from most Christians. Christians also have a responsibility to work with others in transforming government policies, corporate practices, and the behaviour of others who are consumers and major players in the world's market economy (chap. 7). The silence of many churches and individual Christians on these issues is, from God's perspective, ethically unacceptable.

Conclusion

While technological advances, economic cost-benefit analyses, and political policies hold hope for resolving climate change problems, the wisdom of the religions followed by individuals in Canada's multicultural and multireligious mosaic has much to contribute in our response to the challenges of climate change. At the most basic level, these religions have the potential to change the attitudes of many people. Just as important are the effects their many beliefs about nature and the social order can have in providing a humane, ethical, and spiritual underpinning to what is an otherwise soulless "independent" market economy. These various perceptions of ethical and spiritual limits can work with public and private policies in attempting to control the market economy and its complex interaction with climate change. At the very least, Hurka's ethical framework and these religious traditions can give individuals teachings and practices that will allow them to make helpful decisions when faced with the hard choices posed by climate change.

References

Al Faruqi, I. (1989). Islam Ethics. In *World Religions and Global Ethics,* ed. S. Cromwell Crawford. New York: Paragon House, 212–37.

Barbour, I. (1980). *Technology, Environment and Human Values.* New York: Praeger.

Brown, J.E. (1973). Modes of Contemplation Through Action: North American Indians. *Main Currents in Modern Thought* 30: 192–97.

Callicott, J.B. (1989). *In Defense of the Land Ethic: Essays in Environmental Philosophy.* Albany: State University of New York Press.

Callicott, J.B., and R.T. Ames. (1989). *Nature in Asian Traditions and Thought.* Albany: State University of New York Press.

Chan, Wing-tsit, trans. and comp. (1969). *A Source Book in Chinese Philosophy.* Princeton, NJ: Princeton University Press.

Cook, F.H. (1989). The Jewel-Net of India. In *Nature and the Asian Traditions of Thought: Essays in Environmental Philosophy,* ed. J.B. Callicott, and R.T. Ames. Albany: State University of New York Press, 213–29.

Cragg, K. (1977). *The House of Islam.* Belmont, CA: Dickenson.

Crawford, S. Cromwell. (1989). Hindu ethics for modern life. In *World Religions and Global Ethics,* ed. S.C. Crawford. New York: Paragon House, 5–35.

Gross, R. (1997). Towards a Buddhist Environmental Ethic. *Journal of the American Academy of Religion* 65(2): 333–54.

Hurka, T. (1993). Ethical Principles. In *Ethics and Climate Change: The Greenhouse Effect,* ed. H. Coward and T. Hurka. Waterloo: Wilfrid Laurier University Press, 23–28.

Keller, C. (1995). A Christian Response to the Population Apocalypse. In *Population Consumption and Ecology,* ed. H. Coward. Albany: State University New York Press, 109–21.

Leopold, A. (1970). *A Sand County Almanac.* New York: Oxford University Press.

Levy, S. (1995). Judaism and the Environment. In *Population, Consumption and the Environment,* ed. H. Coward. Albany: State University of New York Press, 73–107.

McDonough, S. (1984). *Muslim Ethics: A Comparative Study of the Ethical Thought of Sayyid Ahmad Khan and Mawlana Mawdudi.* Waterloo: Wilfrid Laurier University Press.

Nash, J.A. (1991). *Loving Nature: Ecological Integrity and Christian Responsibility.* Nashville: Abingdon Press.

Nasr, S.H. (1987). The Cosmos and the Natural Orabi. In *Islamic Spirituality: Foundations,* ed. K. Pachauri. New York: Crossroad, 345–60.

Northwest Earth Institute. (2000). *Discussion Course on Voluntary Simplicity.* Portland, OR: Northwest Earth Institute.

Rawls, J. (1971). *A Theory of Justice.* Cambridge, UK: Harvard University Press.

Ruether, R.R. (1992). *Gaia and God: An Ecofeminist Theology of Earth Healing.* New York: Harper Collins.

Schorsch, I. (1992). Trees for Life. *The Melton Journal* 25 (Spring): 1–10.

Segal, E. (1998). Judaism and Ecology. *The Jewish Star,* 26 May 1998.

Sewid-Smith, Daisy. (1995). Aboriginal Spirituality, Population, and the Environment. In *Population, Consumption and the Environment,* ed. H. Coward. Albany: State University of New York Press, 73–71.

Shapiro, R.M. (1989). Blessing and Curse: Toward a Liberal Jewish Ethic. In *World Religions and Global Ethics,* ed. S.C. Crawford. New York: Paragon House, 155–87.

Svedin, U. (1995). Christopher Columbus' Situation and the Challenge of Understanding Today's Global Environmental Issues. *European Review* 3(1): 93–101.

Wei-ming, Tu. (1989). The Continuity of Being: Chinese Visions of Nature. In *Nature in Asian Traditions of Thought,* ed. J. Baird Callicott and Roger T. Ames. Albany: State University of New York Press, 67–78.

World Commission on Environment and Development. (1987). *Our Common Future.* Oxford: Oxford University Press.

Concluding Remarks

Andrew J. Weaver

CANADIANS ARE OBSESSED WITH WEATHER, its variability, and its effect on everything we do. We are also greatly concerned with our climate, although the difference between weather and climate is often not well understood. By definition, climate is the statistics of weather including, for example, its mean and variance. A torrential downpour is an individual weather event, whereas the likelihood of its occurrence in any given year is an aspect of our climate that is derived from the long-term averages of many individual weather events.

When we discuss climate change, we are discussing the change in the statistics of weather. The term *global warming* refers to the increase in the Earth's global mean temperature as a direct consequence of the increased atmospheric loading of greenhouse gases arising from fossil-fuel combustion. The basic physics of global warming are not complicated and have long been established in the scientific literature. The interesting science questions of the present concern how global warming will change the regional statistics of weather, and the so-called feedback mechanisms that may affect the magnitude of these changes.

It is clear that projections of future global warming share common themes. There will be amplified warming at high latitudes relative to the tropical latitudes, in the northern relative to southern hemisphere, in the winter relative to summer, over land relative to ocean, and at night relative to day. There will be an increase in mid- and high-latitude precipitation, especially in the winter and spring, although with warmer temperatures, and hence later winter freezes and earlier spring thaws, one might expect a large component of this to be in the form of rain rather than snow. There will be an increase in extreme precipitation events. Despite this, there will be an increased likelihood of summer drought. There will be a large-scale retreat of most of the world's glaciers with less short-term impact on the Greenland and Antarctic ice sheets. Sea ice in the Arctic will melt back dramatically in the summer.

Conveying the significance of global warming to the public is a difficult task for scientists. In the summer of 2002, for example, much media attention was given to the torrential rainfalls and flooding in Europe and India. At the same time, the Canadian prairie farmers were suffering the worst drought on record. The dilemma is that when asked, "Are these events caused by global warming?" the scientist must respond with a long discussion of weather, climate, and the relationship between the statistics of weather and climate. The media simply want a yes-or-no answer. We will never be able to say that a particular weather event is caused by global warming. Rather, what science can offer is a quantification of the change in the likelihood of such an event. For example, under global warming an expected 20-year return precipitation event in the present climate may become a 10-year return event and, after a few more decades, a five-year event.

Nevertheless, it is clear that climate change is upon us. The reality is that even if we dramatically cut fossil-fuel emissions today, we have warming in store for centuries due to the slow response time of the climate system. We must ask ourselves a more relevant policy question: what do we as a collective society deem to be an acceptable level of climate change?

There is a very real danger that Canadians will believe Kyoto is the answer to climate change. Kyoto is only a small, yet important, first step on our road to a sustainable future no longer dependent on fossil fuels. Even if all countries meet their Kyoto targets, climate will be negligibly affected over the rest of this century—but what do we have without the Kyoto Protocol? We have nothing but unbounded growth in emissions with no end in sight.

Throughout the history of humans, weather and its climate have influenced the growth and fall of civilizations and the livelihood and economic well being of their people. What is different between then and now? Technology. In the past, humans did not have the economic or technological wherewithal to rapidly adapt to the challenges posed by changes in the statistics of weather. Today, large-scale irrigation, fertilization, and land management techniques have substantially improved our adaptive strategies. The 2002 prairie drought may be the worst in our meteorological record, but it certainly pales in comparison to the dust-bowl years earlier this century in terms of its effect on Canadian society.

So if we can adapt to climate, why should we worry? The answer is simple. Large parts of the world have neither the technological nor the economic ability to adapt to a changing climate. Couple this with the fact that the problem was caused by industrialized nations, and you have the seeds of discontent. These seeds could grow to resentment and hostility or, if we take appropriate international steps to mitigate climate change and assist non-industrialized nations in adapting to its effects, these same seeds could blossom into a mechanism for creating global stability. In short, dealing with climate change is about dealing with domestic and global security.

Dealing with climate change has immediate co-benefits. Severe smog conditions in Toronto, Montreal, the Maritimes, and southeast Asia have been in the news of late. Yet smog, like carbon dioxide, is a byproduct of fossil-fuel consumption.

The climate change we are in store for over the next few centuries will be larger and occur faster than at any time in the last 10,000 years. While our pace of technological advance has historically been fast, it must remain faster than the pace at which climate will change in the future. Kyoto, and the required post-Kyoto agreements, are the necessary incentives to ensure that these technological targets are met. Along with such steps by governments, individuals need to examine the hard choices that will lessen the impact of climate change on future generations.

❋

About the Authors

Brad Bass has been a member of Environment Canada's Adaptation and Impacts Research Group (AIR) since 1994. His research interests broadly include complexity, using ecological technologies to adapt to climate change and the impacts of climate change on the energy sector. More specifically, he is developing the COBWEB software to explore how a system of individual agents adapts to change and the emergence of different attractors in a complex system. He has been the Environment Canada lead for Green Roofs, collaborating with many other partners on research projects related to energy efficiency, the urban heat island, and stormwater runoff. Currently, he is working with other faculty at the University of Toronto to assess how green roofs and other components of the urban forest can be integrated with other measures to reduce energy consumption at a neighbourhood scale. In collaboration with partners at the University of Regina, he has examined the impact of climate change on the energy sector in the Toronto-Niagara Region and has developed a regional-scale energy model for this type of analysis.

James P. Bruce, OC, FRSC—Jim Bruce is a senior associate of Global Change Strategies International, Inc., and Canadian policy representative of the Soil and Water Conservation Society. His more than 40-year career has been in the fields of meteorology, climate, water resources, disaster mitigation, and environment as research scientist, and later in senior executive positions within the Canadian government and UN organizations. From 1986 to 1989, he was director of technical cooperation and acting deputy secretary-general of the World Meteorological Organization, Geneva. In the 1990s, he completed terms as co-chair of the Intergovernmental Panel on

Climate Change (IPCC) Working Group III on economics, and as chair of the Canadian Climate Program Board and chair of the UN's Scientific and Technical Committee for the International Decade for Natural Disaster Reduction. He is now vice-chair of the Board of the International Institute for Sustainable Development. He has been made an Officer of the Order of Canada and holds honorary doctorates from the University of Waterloo and McMaster University. Recent awards include the Massey Medal of the Canadian Geographical Society and the IMO Prize of the World Meteorological Organization for "exceptional world-wide contributions in meteorology and hydrology."

Stewart J. Cohen is a scientist with the Adaptation and Impacts Research Group, Meteorological Service of Canada of Environment Canada, and an adjunct professor with the Institute for Resources, Environment and Sustainability, University of British Columbia. He received his PhD from the University of Illinois in 1981. He works primarily on the regional impacts of climate and climate change, and has organized case studies throughout Canada, including the 1990–1997 *Mackenzie Basin Impact Study*, published by Environment Canada (1997). He was a lead author of the chapter on North America in the Intergovernmental Panel on Climate Change (IPCC) Third Assessment Report volume *Climate Change 2001: Impacts, Adaptation and Vulnerability* (2001). He currently serves on the editorial boards of *Climatic Change* and *Integrated Assessment* (formerly *Environmental Modelling and Assessment*), and is serving as science director of the British Columbia node of the Canadian-Climate Impacts and Adaptation Research Network (C-CAIRN BC).

Harold Coward is past director of the Centre for Studies in Religion and Society and professor of history at the University of Victoria. He received his PhD from McMaster University. His main fields are comparative religion and environmental ethics. He has served as an executive member of the board of the Canadian Global Change Program. He is a Fellow of the Royal Society of Canada. He has been the recipient of numerous research grants from SSHRC and the Ford Foundation. He has been a visiting Fellow at Banaras Hindu University and the Institute for Advanced Studies in the Humanities, Edinburgh University. He has written 65 articles and is author/editor of 36 books, including: *Hindu Ethics* (1988); *The Philosophy of the Grammarians* (1990); *Derrida and Indian Philosophy* (1990); *Ethics and Climate Change: The Greenhouse Effect* (1993); *Population, Consumption and the Environment* (1995); *Visions of a New Earth* (2000); and *Just Fish: Ethics and Canadian Marine Fisheries* (2000).

David Etkin has been with Environment Canada since 1977. During his career, he has been a weather forecaster in Nova Scotia and Ontario, taught meteorology to new forecasters, and done applied research in the Arctic and Industrial Climatology Divisions of the Canadian Climate Centre. In 1993 he joined the Adaptation and Impacts Research Group of the Meteorological Service of Canada, specializing in the interdisciplinary study of natural hazards and disasters. Since 1996 he has worked at the University of Toronto, with the Institute for Environmental Studies. He has contributed to several national and international natural hazard projects, including the 2nd US national assessment, and is currently PI of the Canadian National Assessment of Natural Hazards. He has authored 21 peer-reviewed publications, 27 reports, and edited one book on natural hazard related issues.

Brenda Jones is a PhD student in the Department of Geography at the University of Waterloo (UW). She has worked as a research associate for the Adaptation and Impacts Research Group since 2000 at their UW location. Her research interests include hazards and the spatial representation and analysis of vulnerability in the urban context.

Jacinthe Lacroix works as a natural hazards specialist at the Public Safety Branch of the Quebec Ministry of Public Security, where she conducts impact studies on meteorological and hydrological hazards and on climate change impacts and adaptation. She is currently working on a PhD in hydroclimatology, on the subject of "Climate Change and Floods in Southern Quebec." She is president of the Quebec Climatology Association (ACLIQ).

Nigel Livingston is the director of the Centre for Forest Biology and a professor in the Department of Biology, University of Victoria. He received his PhD from the University of British Columbia in 1986. His research expertise covers whole plant physiology, micrometeorology, and environmental instrumentation. His research focuses on the physical and biological interaction between plants and the atmosphere with an emphasis on the mechanisms that determine carbon uptake and water loss by conifers. He serves on or has served on a number of national and international committees, including the International Standards Organization and the Expert Committee on the Regulations Relating to Genetically Modified Trees.

Steve Lonergan is a professor in the Department of Geography, University of Victoria. He received his PhD from the University of Pennsylvania in 1981. He works on issues of environment and security, water in the Middle

East, and the consequences of climate change. He is the author of *Watershed: The Role of Water in the Israeli-Palestinian Conflict* (with D. Brooks, 1994). He is also editor of the policy briefing series Aviso and is past director of the Global Environmental Change and Human Security project (GECHS) for the International Human Dimensions Programme on Global Environmental Change.

Murray Love is a research associate at the Institute for Integrated Energy Systems, specializing in computer modeling of renewable energy systems. He received his bachelors and masters degrees from the department of Mechanical Engineering at the University of Victoria. His research focuses on the technological, economic, and political implications of society's choices of energy technologies, and he aims to shed light on some less well-understood aspects of renewable energy systems, such as their land area and energy storage requirements. He hopes that his findings will enable citizens to make informed decisions about the future of our energy system.

Alastair Lucas, QC, is chair of Natural Resources Law and adjunct professor of environmental science at the University of Calgary and director of the University of Calgary–Latin American Energy Organization Energy and Environmental Law Project. He received an LLB from the University of Alberta in 1966 and an LLM from the University of British Columbia in 1967. His research is in domestic and international energy and environmental law. He is co-editor of Butterworths' *Canadian Environmental Law, and Emond-Montgomery's Environmental Law and Policy,* co-author of *Oil and Gas Law in Canada,* and author of various articles on energy, environment, and natural resources law. He is a legal advisor to the North American Commission for Environmental Cooperation.

Gerard F. McLean is the past director of the Institute for Integrated Energy Systems and associate professor of mechanical engineering at the University of Victoria. He received his bachelors, masters and PhD degrees from the Department of Systems Design Engineering at the University of Waterloo, has worked in the oil and gas industry, and has been involved in start-up ventures to commercialize novel data acquisition systems. He has a broad background in engineering design with applications including bicycle components and testing, embedded systems for structural monitoring, machine vision, automated photogrammetry, and assistive devices for people with disabilities. He has extensive background in PEM fuel-cell systems design, with numerous inventions in stack topology, manufacturing, and manifolding. He is committed to discovering feasible paths to sustainable energy through the development of new technologies and policy tools, and is committed to public education and awareness of energy issues. His current research focuses on micro-structured fuel cells and historical energy patterns.

Brian Mills is an applied climatologist with the Adaptation and Impacts Research Group, Meteorological Service of Environment Canada, based at the Faculty of Environmental Studies, University of Waterloo. Brian's research is presently focused on understanding the sensitivity and adaptability of road transportation and urban water management practices to weather and climate.

Doug Russell is the president of Global Change Strategies International Co. He has over 25 years experience in the public and private sectors dealing with international and domestic policy development related to broad-scale environmental issues. His recent work includes a corporate climate change strategy for the Shell Group of companies; a 2001 study for the Pew Center on Global Climate Change on how multinational corporations are setting greenhouse gas reduction targets; a study on issues arising from the establishment of a North American system for emissions trading; and experience with capacity-building initiatives related to climate change in the Caribbean, Argentina, Nigeria, and China. He has also developed and delivered numerous presentations and briefings on the Kyoto Protocol and the emergence of carbon emissions as a commodity. Clients have ranged from large corporations to developed and developing country governments to environmental non-government organizations. Prior to moving to the private sector, Mr. Russell co-headed Canada's negotiating delegation to the UN Framework Convention on Climate Change.

Daniel Scott is an assistant professor in geography at the Faculty of Environmental Studies, University of Waterloo. Dr. Scott is the author of over 30 publications related to climate change impacts and adaptation, focusing on Canada's national parks and the tourism industry. He was a contributing author to the UN Intergovernmental Panel on Climate Change Third Assessment Report, and is a co-chair of the International Society of Biometeorology–Commission on Climate Tourism and Recreation.

Gordon Smith is director of the Centre for Global Studies at the University of Victoria. He has a PhD in political science from MIT. His last position in the Canadian government was as deputy minister of Foreign Affairs and the prime minister's personal representative for the G7 and G8 Summits. He has also occupied a number of other senior positions at the deputy minister and ambassadorial level. One of his initiatives in Ottawa was the creation of a Global Issues Bureau in the Department of Foreign Affairs and International Trade. He is chairman of the International Development Research Centre (IDRC) as well as being chairman of the International Network on Bamboo and Rattan (INBAR). He is also director of the Canadian Global Change Program, he chairs the board of the Canadian Institute for Climate Studies, and he is a senior adviser to the rector of UPEACE.

G. Cornelis van Kooten is currently the Canada Research Chair in Environmental Studies and Climate Change at the University of Victoria. Before that he was professor of Natural Resource Economics at the University of Nevada at Reno and a professor in the Faculty of Agricultural Sciences and in the Faculty of Forestry at the University of British Columbia. He also holds a part-time appointment in the Department of Social Sciences at Wageningen University. He has published more than 100 peer-reviewed journal articles and seven books, including *The Economics of Nature* (Blackwell, 2000) with Erwin Bulte. Much of his research has focused on forestry, agriculture, and climate change.

David G. Victor is an adjunct senior Fellow of the Council of Foreign Relations. He is an expert on the effects of science and technology on foreign affairs. He has served as senior Fellow and director on the Program on Energy and Sustainable Development at Stanford University; Robert W. Johnson, Jr., Fellow for Science and Technology, Council on Foreign Relations; research scholar with the Project on "Environmentally Compatible Energy Strategies" with the International Institute for Applied Systems Analysis; and co-leader of the Project on Implementation and Effectiveness of International Environmental Commitments. His PhD is from the Massachusetts Institute of Technology.

Andrew J. Weaver is a professor and Canada Research Chair in Atmospheric Science in the School of Earth and Ocean Sciences, University of Victoria. He joined UVic in 1992, having spent three years as a Natural Sciences and Engineering Research Council (NSERC) University Research Fellow in the Department of Atmospheric and Oceanic Sciences at McGill University. He has written over 120 peer-reviewed papers in climate, meteorology, oceanography, earth science, policy, and education journals. He was involved as a lead author in the United Nations Intergovernmental Panel on Climate Change (IPCC) second and third scientific assessments of climate change, and is involved in the planning of the IPCC fourth scientific assessment to appear in 2007. He presently serves on the United Nations World Climate Research Program Working Group on Coupled Modelling and the United States National Academy of Sciences (NAS) Climate Research Committee, as well as the NAS Panel on Climate Feedbacks. He has served on numerous other national and international committees over the last decade and is an editor of the *Journal of Climate*. In 1997, he was awarded the NSERC E.W.R. Steacie Memorial Fellowship; in 2001 he was elected Fellow of the Royal Society of Canada; and in 2002, he received a Killam Research Fellowship and a CIAR Young Explorers award as one of the top 20 scientists in Canada under the age of 40, and he was selected as one of the 25 Power Thinkers in British Columbia by *BC Business Magazine*. In 2003 he was selected as one of the top five Canadian Scientists by *Time Magazine Canada*.

Jan Zwicky is an associate professor of philosophy at the University of Victoria. She received her PhD from the University of Toronto in 1981, and has taught at a number of North American universities since then. Her areas of philosophic interest include metaphilosophy, environmental philosophy, and early Greek philosophy. *Wisdom & Metaphor* (2003) is concerned primarily with metaphysical questions and the importance of metaphorical insight to the practice of philosophy. A recent collection of poetry, *Songs for Relinquishing the Earth*, won the Governor General's Award for poetry in 1999.

Index